Textbook of Plant Taxonomy

Textbook of Plant Taxonomy

– Author –
Mukul Tomar Mitra

Wordpen Academics
Publishing Science That Matter

ISBN: 978-93-90438-26-6 (Hardback)

Published by : **WORDPEN ACADEMICS**
4736/23, Ansari Road, Darya Ganj,
New Delhi – 110 002
Phone: +91-011-35004336
E-mail: info@wordpenacademics.com

Printed at **: Replika Press Pvt. Ltd.**

Exclusive Distributor: ***Bio-Green Books, New Delhi***

Preface

Taxonomy is the method by which scientists, conservationists, and naturalists classify and organize the vast diversity of living things on this planet in an effort to understand the evolutionary relationships between them. Modern taxonomy originated in the mid-1700s when Swedish-born Carolus Linnaeus (also known as Carl Linnaeus or Carl von Linné) published his multi-volume *Systema naturae*, outlining his new and revolutionary method for classifying and, especially, naming living organisms. Prior to Linnaeus, all described species were given long, complex names that provided much more information than was needed and were clumsy to use. Linnaeus took a different approach: he reduced every single described species to a two-part, Latinized name known as the "binomial" name. Thus, through the Linnaean system a species such as the dog rose changed from long, unwieldy names such as *Rosa sylvestris inodora seu canina* and *Rosa sylvestra alba cum rubore, folio glabro* to the shorter, easier to use *Rosa canina*. This facilitated the naming of species that, with the massive influx of new specimens from newly explored regions of Africa, Asia, and the Americas, was in need of a more efficient and usable system.

Although trained in the field of medicine, botany and classification were the true passions of Linnaeus and he actively explored northern Europe and described and named hundreds of new plant species during his lifetime. As well, Linnaeus spent a great deal of time describing and naming new plant specimens that were sent to him from around the world by other botanists, including from the newly explored regions of the New World. Linnaeus classified this multitude of new plant species based upon their reproductive structures, a method which is still largely in use today. In fact, the majority of the species described by Linnaeus are still recognized today, indicating how far ahead of his time he truly was. Although somewhat rudimentary by today's standards, Linnaeus' methods of describing species in such a way as to represent the relationships between them changed the face of taxonomy and allowed biologists to better understand the complex natural world around us. Plants, and indeed all organisms, are classified in a hierarchical system that attempts to illustrate the evolutionary relationships between the various groupings within the hierarchy. This concept of relatedness forms the backbone of modern classification schemes. Scientists who attempt to classify organisms and place them within an evolutionary framework are called Taxonomists, the most famous of which would be Linnaeus himself. At the broadest level, all organisms on the planet are classified into 5 Kingdoms: Animalia (animals), Plantae (plants, some multicellular algae), Fungi (fungi), Monera (prokaryotic bacteria), and Protista (eukaryotic bacteria, most algae, etc.), representing the most ancient branches of the evolutionary "tree of life." Organisms in any given Kingdom may be separated from organisms in any other Kingdom by many hundreds of millions, if not billions, of years of evolution. Historically, all organisms known were grouped into only two Kingdoms: organisms that had finite growth, moved, and ate were grouped into the Kingdom Animalia, while organisms that had indefinite growth, didn't move, and didn't eat were grouped into the Kingdom Plantae. Of course, as science progressed, it became increasingly evident that such a simplistic approach to taxonomy was ineffective and many species were found that did not fit either grouping particularly well. The proposal to move to an eight-Kingdom system suggests that our current classification system, with its five Kingdoms, may yet change again as our understanding of the diversity of organisms around us continues to grow.

This book on Plant Taxonomy has been structured to cover every aspect of the subject. The text provides a meaningful, realistic context for learning all of the core material that students must master as per their syllabi. It gives them with accurate modern and streamlined content with illustrations to make the material easier to understand.

Author

Contents

Chapter 1

Plant Organs

Roots

The root apical meristem, or root apex, is a small region at the tip of a root in which all cells are capable of repeated division and from which all primary root tissues are derived. The root apex is protected as it passes through the soil by an outer region of living parenchyma cells called the root cap. As the cells of the root cap are destroyed and sloughed off, new parenchyma cells are added by a special internal layer of meristematic cells called the calyptrogen. Root hairs also begin to develop as simple extensions of protodermal cells near the root apex. They greatly increase the surface area of the root and facilitate the absorption of water and minerals from the soil.

Along the longitudinal axis of a root, beginning with the root cap and leading away from the root tip, there are five distinct zones in which certain specific growth patterns dominate: cell division, cell elongation, primary tissue maturation, mature primary tissues, and secondary tissue growth (the latter is found in woody roots—i.e., those of perennial dicotyledons). There is a gradual transition between these regions.

The region of cell division includes the apical meristem and the primary meristems—the protoderm, ground meristem, and procambium—derived from the apical meristem. As is generally true of nonmeristematic regions elsewhere in the plant body, root length in the second region is increased as a result of cell elongation rather than by cell division. The region of maturation that follows is where the cells differentiate (i.e., change in structure and physiology into cells of a specific type) and where the first primary phloem and xylem, as well as mature root hairs, are clearly seen. The region of mature primary tissues is where the anatomy of the primary body of the root is most obvious and where all the elements of the vascular cylinder, cortex, and epidermis are evident. Finally, in the region of secondary growth, the secondary xylem and phloem as well as the periderm add girth to the plant.

There are many individual vascular strands (or vascular bundles) in the primary body of the stem, and they all converge into a single central vascular cylinder in the root, forming a continuous system of vascular tissue from the root tips to the leaves. At the centre of the vascular cylinder of most roots is a solid, fluted (or ridged) core of primary xylem

The vascular cylinder of the root is surrounded by a layer of parenchymatous pericycle cells. As the root ages, many of these cells become fibres, particularly in monocotyledons and many herbaceous dicotyledons. As defined above, a characteristic feature of parenchyma cells is their ability to differentiate into cells of a different type under appropriate conditions. The parenchyma cells of the pericycle, then, can be considered meristematic in that they give rise to new lateral meristems and lateral roots. In woody roots the vascular cambium (the lateral meristem that gives rise to secondary phloem and secondary xylem) originates in the pericycle as well as in the procambium; the procambium is the primary meristematic tissue between the primary phloem and xylem. The first cork cambium is a lateral meristem that arises in the pericycle; the successive cork cambia arise in the secondary phloem. Because lateral roots are initiated in the pericycle and grow out through the cortex and epidermis, they are said to have an internal, or endogenous, origin, in contrast to the external, or exogenous, origin of leaves and the apical meristem of stems.

Ground tissue called the cortex surrounds the vascular cylinder and pericycle. The cortex of roots generally consists of parenchyma cells with large intercellular air spaces. The endodermis (the innermost layer of the cortex adjacent to the pericycle) is composed of closely packed cells that have within their walls Casparian strips, water-impermeable deposits of suberin that regulate water and mineral uptake by the roots.

The cortex is surrounded by the dermal system consisting of a single layer of epidermal cells.

The few variations that occur in root anatomy are mainly found among the monocotyledons. The roots of monocotyledons lack secondary growth. Monocotyledons also generally have a parenchymatous pith in the centre of the vascular cylinder and fibres or sclereids, or both, in the cortex; and extensive well-developed pericyclic fibres. Orchids have a multiple-layered epidermis called a velamen, which consists of nonliving compact cells with lignified strips of secondary walls. These cells provide support, prevent water loss, and assist the plant in absorbing water. When dry the orchid root appears white, and when wet the root appears green because the cells of the velamen absorb water, become translucent, and reveal the green cortical cells.

Stems

The shoot apical meristem and the primary meristems lie at the apex of the shoot and give rise to the primary tissues of the stem. The shoot apical meristem produces leaves and axillary buds exogenously; as a result, the epidermis of stems and leaves is continuous. (In contrast, as mentioned above, the lateral roots are produced endogenously, and the dermal system of the lateral roots is discontinuous with that of the parent root.)

The stem has growth periods similar to those of the root, but longitudinal regions are not as obvious as in the root until the nodes become differentiated and internode lengths increase. Elongation of internodes involves many cell divisions and is followed by cell elongation. At this point, growth in thickness involves some radial cell division and cell enlargement.

The primary tissue systems appear after internode elongation. The procambium differentiates as a basically continuous hollow cylinder or discrete procambial strands, which differentiate into primary xylem and phloem. The ground tissue that lies outside the procambial cylinder is the cortex, and that within is the pith. Ground tissue called the interfascicular parenchyma lies between the procambial strands and remains continuous with the cortex and pith. As the vascular tissue grows, xylem and phloem develop, the vascular bundles mature, the single-layered epidermis differentiates as epidermal cells, trichomes, and a few stomata, and the parenchymatous pith may develop as collenchyma or contain sclereids or fibres or both; unequal pith proliferation and expansion produces the flattened stems (pads) of prickly-pear cacti (*Opuntia;* Cactaceae). The parenchymatous cortex also may develop some collenchyma, sclereids, or fibres; unequal growth and expansion of the cortex produces the cladodes of epiphytic cacti (e.g., night-blooming cereus, *Selenicereus;* Cactaceae). In most aquatic angiosperms, the parenchymatous cortex contains large intercellular spaces. As a rule, angiosperm stems have no endodermis or definable pericycle.

The most common arrangement of the primary xylem and phloem is called a collateral bundle; the outer portion of the procambium (adjacent to the cortex) becomes phloem, and the inner portion (adjacent to the pith) becomes xylem. In a bicollateral bundle, the phloem is both outside and inside the xylem, as in Solanaceae (the potato family) and Cucurbitaceae (the cucumber family). In the monocots, the phloem may surround the xylem, or the xylem may surround the phloem.

The vascular bundles of the stem are continuous not only with the primary vascular system of the root but also with the vascular bundles of the leaves. At each node, one or more longitudinal stem bundles enter the base of the leaf as leaf traces, connecting the vascular system of the stem with that of the leaf. The point at which the stem bundle diverges from the vascular cylinder toward the leaf is a leaf gap, called a lacuna. The number of lacunae varies among angiosperm groups and remains a characteristic for classifying the various species.

Several leaves in a line along the stem have common stem bundles. In some species all stem bundles and their associated leaf traces are interconnected, while in others each stem bundle and the associated leaf trace remains laterally independent of the others. An arrangement of two trace leaves and a single lacuna is found among several primitive angiosperm families and throughout the gymnosperms and is the organisation from which other nodal patterns are derived.

In woody dicots, the vascular cambium is formed in parts that grow toward each other and unite. Each vascular bundle develops a meristematic area of growth from an undifferentiated (parenchymatous) layer of cells between the primary xylem and primary phloem, called a fascicular cambium. This meristematic area spreads laterally from each bundle and eventually becomes continuous, forming a complete vascular cambium.

The arborescent (treelike) stems of monocotyledons have a different growth pattern and anatomy from dicotyledons. Scattered throughout the ground tissue are vascular bundles with no fascicular cambia and no definable pith or cortex. Secondary growth, when it occurs, is different because a secondary thickening meristem forms under the epidermis. This secondary thickening meristem produces secondary parenchyma (conjunctive tissue) to the inside, and then secondary vascular bundles develop within this conjunctive tissue. Thus, there are no rings of secondary xylem or secondary phloem as in woody dicotyledons.

Many arborescent monocots have only massive primary growth without secondary growth. This primary growth is derived from a primary-thickening meristem under the leaf bases that is a lateral continuation of the apical meristem. This primary-thickening meristem produces vast amounts of parenchyma to the inside, through which the leaf traces differentiate.

Leaves

Leaves initially arise from cell divisions in the shoot apical meristem. A slight bulge (a leaf buttress) is produced, which in dicots continues to grow and elongate to form a leaf primordium. (Stipules, if present, appear as two small protuberances.) Marginal and submarginal meristems on opposite flanks of the primordium initiate leaf-blade formation. Differences in the local activity of marginal meristems cause the lobed shapes of simple leaves and the leaflets in compound leaves. An increase in width and in the number of cell layers is brought about by marginal meristems. Subsequent expansion and increase in length is achieved by cell division and the general enlargement of cells throughout the blade.

Leaf growth is determinate; the tip matures first, and maturation then progresses toward the base, after which the leaf cells cease to grow and divide. (Stem growth is generally indeterminate since the meristems are active indefinitely.) The petiole, when present, and the leaf base become thickened, and often the latter expands laterally and fully or partially encloses the stem. Soon after the cells of the marginal meristems begin to divide, procambial strands differentiate into the leaf from the stem bundles to form the midvein, or midrib. The smaller lateral veins of the leaf are initiated near the leaf tip; subsequent major lateral veins are initiated sequentially toward the base, following the overall pattern of leaf development. A major lateral vein may have one or more orders of smaller veins, which also are initiated in size from larger to smaller. This results in the netlike venation patterns characteristic of dicotyledonous leaves.

The anatomy of a mature dicot leaf generally reflects the habitat, especially the availability of water. Mesomorphic leaves are adapted to conditions of abundant water and relatively humid conditions; xeromorphic leaves are adapted to dry conditions with relatively low humidity; and hydromorphic leaves are adapted to aquatic situations, either submerged or in standing water. Mesomorphic leaves (the most common type) are characteristic of crop plants, such as tomatoes and soybeans. Their veins (vascular bundles) permeate the ground tissue of the dermal system—a single layer of epidermal cells with interspersed guard cells. The ground tissue system, the mesophyll, is divided into two regions: the palisade parenchyma, located beneath the upper epidermis and composed of columnar cells oriented perpendicular to the leaf surface, and spongy parenchyma, located in the lower part of the leaf and composed of irregularly shaped cells. The veins contain primary xylem and phloem and are enclosed by a layer of parenchyma called the bundle sheath. Only the midvein and some large lateral veins have any secondary growth.

The anatomy of mesomorphic leaves is designed to function optimally for water uptake and gas exchange in photosynthesis under mesic (moist) conditions. The spongy mesophyll with irregularly shaped cells provides increased surface area internally, while the elongate palisade cells provide optimal exposure of chloroplasts to light. The anatomy of hydromorphic leaves is simplified: the cuticle is thin or lost; the guard cells are raised and are found only on the upper surface in floating leaves (they are lost in most submerged leaves); the mesophyll contains aerenchyma (an adaptation to promote water loss) and little or no collenchyma or sclerenchyma; and the vascular system (particularly the water-conducting element of the vascular system, the xylem) is only weakly developed since the water provides much of the mechanical support to the plant normally provided by the xylem.

The abundance of water means that there is no need for mechanisms that prevent water loss and little need for additional supports. The leaves generally become large and thin, and the reduction or loss of cuticle, vascular tissue, and ground tissue (mesophyll) permits the rapid loss of water vapour (transpiration). The guard cells on the upper surface of floating leaves also monitor the rate of water loss through the central stomata. Such plants may wilt if the turgor (water) pressure is reduced. Water lilies (Nymphaeaceae) and rice crops contain hydromorphic leaves.

Xeromorphic adaptations to arid conditions are quite varied and tend to prevent water loss during periods when water is limited and must be conserved by the plant. There are many modifications limiting transpiration: two examples are a multilayered epidermis covered by thick layers of epicuticular wax or mucilages secreted into stomates; another is dense mats of trichomes on both surfaces of the leaf and guard cells and stomata sunken into the lower surface and often lined with numerous trichomes, which trap moisture, thereby inhibiting total water loss. Mesophyll modifications provide a means of storing water. Most xeromorphic leaves have a high volume-to-surface ratio—i.e., they are small and compact. In addition, many are fleshy and often oval to round in shape.

The development of monocotyledonous leaves after initiation of cell division on the shoot apical meristem is different from that of dicotyledons and results in leaves with different morphologies from those of dicotyledons. Leaves in monocotyledons have either a radial leaf tip or are expanded in the same plane as the stems instead of at a right angle to the stem, as in dicotyledons. The leaf buttress begins as a ring that encloses the stem. The upper portion of the buttress develops a meristem on the side facing the stem (adaxial meristem). Growth at this adaxial meristem forms the flattened leaf with the radial (cylindrical) leaf tip typical of the monocotyledons. If the adaxial meristem is long-lived, long flat leaves in the same plane as the stem are formed (*Iris;* Iridaceae); if short-lived, flat leaves with short cylindrical tips develop (snake plant, *Sansevieria trifasciata;* Agavaceae). When the radial (topmost aspect of the leaf) is short, the base becomes flattened because the marginal meristems (those on either side of the midvein) continue to expand outward. A monocot leaf grows either radially or along the margins, but not both in the same region. The monocot leaf grows in length from a meristem at its base, which is why it is possible to mow grass and have the leaf blades continue to grow.

The developmental pattern from a basal intercalary meristem has placed constraints on the anatomy of monocot leaves, particularly with respect to venation and the position of stomates. This has produced a leaf anatomy characteristic of the monocots. There is no midvein, and veins are longitudinally parallel. The stomates are in rows between the veins, and the mesophyll is often poorly developed and mostly parenchymatous with scattered bundles of fibres. Thus, most monocot leaves are uniform in appearance and texture. Most of the hydromorphic and xeromorphic modifications found among dicot leaves, however, also occur in monocot leaves in similar environments.

Reproductive Structures

General Features

The broad range of variation in the morphology and structure of nonreproductive (vegetative) organs within the angiosperms has been outlined above. There is a similarly broad range in the morphology and structure of the reproductive organs of the plant. Many vegetative buds sooner or later become flower buds. Flower buds are modified leaves borne on a short axis with very short internodes and no axillary buds. The floral axis has determinate growth, in that at some point it ceases to grow. Flowers, the reproductive tissues of the plant, contain the male and/or female organs. They may terminate short lateral branches or the main axis or both. Flowers may be borne singly (as in the daffodil and *Magnolia*) or in clusters called inflorescences (e.g., bromeliads, snapdragons, and sunflowers). Fruits are derived from the floral parts of the angiospermous plant.

A complete flower is composed of four organs attached to the floral stalk by a receptacle

The sepals, the outermost layer, are usually green, enclose the flower bud, and collectively are called the calyx. Petals are the next layer of floral appendages internal to the calyx; they are generally brightly coloured and collectively are called the corolla. The calyx and corolla together compose the perianth. The sepals and petals are accessory parts or sterile appendages; though they protect the flower buds and attract pollinators, they are not directly involved with sexual reproduction. When the colour and appearance of sepals and petals are similar, as in the tulip tree (*Liriodendron tulipifera*) and Easter lily (*Lilium longiflorum*), the perianth is said to be composed of tepals.

Internal to the corolla are the stamens, spore-producing structures (microsporophylls) that are collectively called the androecium. In most angiosperms, the stamens consist of a slender stalk (the filament) that bears the anther (and pollen sacs), within which the pollen is formed.

Small secretory structures called nectaries are often found at the base of the stamens and provide food rewards for pollinators. In some cases the nectaries coalesce into a nectary or staminal disc. In many cases the staminal disc forms when a whorl of stamens is reduced into a nectiferous disc, and in other cases the staminal disc is actually derived from nectary-producing tissue of the receptacle.

At the centre of the flower are the carpels, collectively called the gynoecium. Carpels are megasporophylls that enclose one or more ovules, each with an egg. After fertilization, the ovule matures into a seed, and the carpel matures into a fruit. Carpels, and thus fruit, are unique to angiosperms.

A complete flower contains all four organs, while an incomplete flower is missing at least one. A bisexual (or "perfect") flower has both stamens and carpels, and a unisexual (or "imperfect") flower either lacks stamens (and is called carpellate) or lacks carpels (and is called staminate). Species with both staminate flowers and carpellate flowers on the same plant (e.g., corn) are monoecious, from the Greek for "one house." Species in which the staminate flowers are on one plant and the carpellate flowers are on another are dioecious, from the Greek for "two houses."

Floral organs are often united or fused: connation is the fusion of similar organs—e.g., the fused petals in the morning glory; adnation is the fusion of different organs—for example, the stamens fused to petals in the mint family (Lamiaceae). The basic floral pattern consists of alternating whorls of organs positioned concentrically: from outside inward, sepals, petals, stamens, and carpels. It is possible in most cases to interpret the flower with respect to missing parts and/or the modification of parts to function as missing parts simply by positional relationships.

In a complete five-merous flower (starting from the outside) there would be a whorl of five sepals, followed by an alternating whorl of five petals, followed by an alternating set of five stamens. In the floral diagram, the midline of each petal is midway between the midlines of two adjacent sepals. Because the whorls alternate, the midline of each stamen of the stamen whorl is between the midlines of two adjacent petals and on the midline of each sepal. When the petals are missing and bracts appear coloured and petaloid as in the *Bougainvillea*, one of the three whorls is missing: there are only two whorls of five organs instead of the three whorls of five organs described above. Because one whorl of the flower is obviously composed of stamens that bear functional pollen and the other whorl is composed of a brightly coloured set of organs that resemble petals one might conclude that the sepals are missing. But examination of positional relationships between the whorls reveals that the midline of each stamen is on the same line as the midline of the organs of the brightly coloured set. Thus, position tells us that the brightly coloured whorl represents a sepal whorl and that the sepals have assumed the function of the missing petals.

The Receptacle

The receptacle is the axis (stem) to which the floral organs are attached. Floral organs are attached either in a low continuous spiral, as is common among primitive angiosperms, or in alternating successive whorls, as is found among most angiosperms.

The peduncle is the stalk of a flower or an inflorescence. When a flower is borne singly, the internode between the receptacle and the bract (the last leaf, often modified and usually smaller than the other leaves) is the peduncle. When the flowers are borne in an inflorescence, the peduncle is the internode between the bract and the inflorescence; the internode between the receptacle of each flower and its underlying bracteole is called a pedicel. Thus, in inflorescences, bracteole is the equivalent of bract, and pedicel is the equivalent of peduncle.

Often the bract subtending an inflorescence is brightly coloured, as in the poinsettia (*Euphorbia pulcherrima;* Euphorbiaceae), or provides protection, as in the woody, boat-shaped bracts in many palms. Bracteoles in the inflorescence of *Bougainvillea* also are brightly coloured to attract

pollinators. In some angiosperms, the receptacle becomes fleshy; in the strawberry, for example, the receptacle is the fleshy edible part of the strawberry and, when eaten by small mammals and birds, aids in seed dispersal. In others, the peduncle or pedicel becomes fleshy; in the cashew (*Anacardium occidentale;* Anacardiaceae), for example, the pedicel is made into a drink in the Neotropics, and it also aids in fruit dispersal of the much smaller cashew nut. In cacti (e.g., prickly pear), the fleshy part of the edible fruit forms from the receptacle and peduncle, and several internodes below that grow up and surround the carpels; this is why there are axillary buds in cacti (areoles) with spines on the fruit surface.

The Calyx

The sepals (collectively called the calyx) most resemble leaves because of their generally green colour. From their base and along most of their length, sepals remain either separate (aposepalous, or polysepalous) or marginally fused (synsepalous), forming a tube with terminal lobes or teeth. The number of calyx lobes equals the number of fused (connate) sepals. The sepals enclose and protect the unopened flower bud. The calyx is commonly persistent and evident when the fruit matures (e.g., persimmon, *Diospyros virginiana;* Ebenaceae), in contrast to the more short-lived petals and stamens. Sepals may be brightly coloured and function as petals when true petals are missing—for example, the virgin's bower (*Clematis;* Ranunculaceae) and the *Bougainvillea.* Petaloid sepals in this case differ from tepals because the first group of stamens are on the same radii as the sepals, indicating the absence of the petals, which would normally be positioned on alternating radii in the next floral whorl.

The Corolla

The petals composing the corolla are typically brightly coloured or white and attract insects and birds for pollination. The number of petals is usually the same as the number of sepals. Floral symmetry is defined by the petals. When the petals of the corolla are of the same size and shape and when they are equidistant from each other, the flower has radial symmetry, and the flower is called regular or actinomorphic (e.g., buttercup, *Ranunculus;* Ranunculaceae). In regular flowers, any line drawn through the centre will divide the flower into two identical halves. When at least one petal of the corolla is different, the flower has bilateral symmetry and is called irregular or zygomorphic (e.g., violets, *Viola;* Violaceae [see photograph]).

The petals of the corolla may be separate, or apopetalous, or marginally fused (fusion of like floral parts is called connation), or sympetalous, for all or part of their length. When joined, they form a tubular corolla with terminal lobes. A tubular corolla may be present in regular flowers (e.g., blueberries, *Vaccinium;* Ericaceae) or irregular flowers (e.g., sage, *Salvia officinalis;* Lamiaceae). Stamens are commonly united to a tubular corolla (fusion of two unlike floral parts is called adnation). A marginally fused (synsepalous) calyx, a marginally fused (sympetalous) corolla, and stamens may fuse to form a cuplike floral tube called a hypanthium that surrounds the carpels, as in cherries (*Prunus;* Rosaceae), for example. Fusion and reduction of flower parts are common and have occurred in many unrelated lineages. Many wind-pollinated angiosperms do not have petals, nor do they have floral parts modified as petals; examples of wind-pollinated species include the amaranth family (Amaranthaceae) and the birch family (Betulaceae).

Petals often bear nectaries that secrete sugar-containing compounds, and petals also produce fragrances to attract pollinators; the fragrance of a rose (*Rosa;* Rosaceae) is derived from the petals. Petals often develop a nectar-containing extension of the tubular corolla, called a spur. This may involve one petal, as in the larkspur (*Delphinium*), or all the petals, as in columbine (*Aquilegia*), both members of the family Ranunculaceae.

The Androecium

Stamens (microsporophylls) are structures that produce pollen in terminal saclike structures (microsporangia) called anthers. The number of stamens comprised by the androecium is sometimes the same as the number of petals, but often the stamens are more numerous or fewer in number than the petals. There are generally two pairs of spore-containing sacs (microsporangia) in a young stamen; during maturation the partition between the adjacent microsporangia of a pair breaks down so that there are only two pollen-containing sacs (one in each lobe of the anther) at the time the stamen releases the pollen.

The least-modified stamens are similar to leaves, with the paired microsporangia located at or near the margins; an example is found in the magnolia family (Magnoliaceae). In more derived stamens, the blade has become modified into a slender stalk, the filament, with the microsporangia at or near the filament apex (the anther). The filaments are very often united with the corolla, but with the anthers either separate, as in primroses (*Primula;* Primulaceae), or united with each other to form a staminal tube that encloses the gynoecium, as in the mallow family (Malvaceae). In thistle (*Cirsium;* Asteraceae) and in other members of the sunflower family, the staminal tube is fused to the lower half of the corolla tube. There are several trends in stamen modification. In many angiosperms, one or more of the stamens is modified and lacks functional anthers. In the most common modification, the filament is expanded to form a petallike blade called a staminode (in the same manner that a sepal forms a petallike blade in some flowers without true petals). The apparent petals in some angiosperm families, such as are found in many members of the pink family (Caryophyllaceae), are staminodial in origin. Wild roses have only five petals and many stamens; however, cultivated roses have been selected for the many apparent petals (but actually staminodes) and few functional stamens. In other cases, stamens have been modified into sterile nectaries involved in pollination. If flowers have a large number of stamens, then the stamens often occur in groups or clusters (fascicles; see photograph), as in the myrtle family (Myrtaceae).

The Gynoecium

The gynoecium is composed of carpels. In more basal families (e.g., Magnoliaceae) the carpels are spirally arranged, and in more advanced families they tend to be arranged in a single whorl. Carpel number varies from one (e.g., bean or legume family [Fabaceae]) to many (e.g., buttercups or raspberries [*Rubus*]).

At the base of a carpel is the ovary, within which develop one or more multicellular structures called ovules that each contain an egg. The upper part of the carpel, the stigma, receives the pollen. A slender stalk called the style often connects the ovary and stigma. The carpels may be separate (apocarpous) or fused together (syncarpous), with the individual carpel walls and cavities (locules) still present. Syncarpy may involve only the ovaries, leaving the styles and stigmas free, as is found in the wood sorrel (*Oxalis*), or it may involve both the ovaries and styles, leaving only the stigmas free, as in the waterleaf (*Hydrophyllum*). The number of carpels in a syncarpous (or compound) ovary generally equals the number of locules (in some cases the inner carpel walls break down, leaving a single locule); in an orange or a grapefruit, for example, the juice sacs are actually trichomes that line the inner carpel walls of each cavity.

The position of the gynoecium with respect to the petals, sepals, and stamens on the floral axis also characterizes the flower. In hypogynous flowers, the perianth and stamens are attached to the receptacle below the gynoecium; the ovary is superior to these organs, and the remaining floral organs arise from below the point of origin of the carpel. In perigynous flowers, a hypanthium (a floral tube formed from the fusion of the stamens, petals, and sepals) is attached to the receptacle below the gynoecium and surrounds the ovary; the ovary is superior, and the free parts of the petals, sepals, and stamens are attached to the rim of the hypanthium. In epigynous flowers, the hypanthium is fused to the gynoecium, and the free parts of the sepals, petals, and stamens appear to be attached to the top of the gynoecium, as in the apple (*Malus;* Rosaceae); the ovary is inferior, and the petals, sepals, and stamens appear to arise from the top of the ovary.

Fruits

Fertilization of an egg within a carpel by a compatible pollen grain results in seed development within the carpel. (Formation of fruit without the fertilization of an egg and subsequent seed development is called parthenocarpy.) A fruit is a ripened ovary (or compound ovary) and any other structure, usually the hypanthium, that ripens and forms a unit with it. This clearly separates a fruit from a vegetable, because a vegetable is derived only from vegetative (nonreproductive) organs. Tomatoes, eggplants, and squashes are fruits, because they are derived from floral parts, whereas carrots, turnips (*Brassica rapa*), and beets are vegetables, because they are roots modified as storage organs in the same manner that potatoes, ginger (*Zingiber officinale*), and onions are modified stems.

Simple fruits develop from a single carpel or from a compound ovary. Aggregate fruits consist of several separate carpels of one apocarpous gynoecium (e.g., raspberries where each unit is a single carpel). Multiple fruits consist of the gynoecia of more than one flower and represent a whole inflorescence, such as the fig and pineapple. Accessory fruits incorporate other flower parts in the development of the mature fruit; for example, the hypanthium is used in forming the pear (*Pyrus;* Rosaceae), and the receptacle becomes part of the prickly pear. The form, texture, and structure of fruits are varied (notably in simple fruits), but most fall within a few categories. The fruit wall, or pericarp, is divided into three regions: the inner layer, or endocarp; the middle layer, or mesocarp; and the outer layer, or exocarp. These regions may be fleshy or dry (sclerified) or any combination of the two, but they are classified as either one or the other.

The three main types of fleshy fruits are berries, drupes, and pomes. Berries are many-seeded simple fruits composed of one carpel or a syncarpous ovary. They are fleshy throughout, but the exocarp ranges in texture: a soft, thin exocarp, as in tomatoes (a berry); a leathery exocarp, as in oranges (a hesperidium); and a somewhat hard exocarp, as in pumpkins (a pepo). In drupes, or stone fruits, there is usually only one seed per carpel or locule. Drupes are fleshy fruits and consist of an inner stony or woody endocarp, which adheres to the seed (peaches, plums, and cherries). The term druplet is used for each unit of aggregate fruit of this type (e.g., raspberries and blackberries). Pomes are fleshy fruits of the rose family (Rosaceae) in which an adnate hypanthium becomes fleshy (apples and pears).

Simple dry fruits are either dehiscent or indehiscent. They are dehiscent if the pericarp splits open at maturity and releases the seeds, or indehiscent if the pericarp remains intact when the fruit is shed from the plant. The three principal types of dehiscent fruits are follicles, legumes, and capsules. Follicles and legumes are each derived from an ovary with a single carpel, and a capsule is derived from several united carpels. As the fruit matures, the pericarp dries and the fruit splits. Whereas follicles split along a single side of the fruit, such as in the milkweeds, columbines, and magnolias, legumes split along both sides, as in the bean family. Capsules have two or more carpels and split open to release their seeds in various ways. They may open longitudinally to expose the seeds within each locule (cavity) or longitudinally along each septum between the locules, as in the agave (*Agave;* Agavaceae). Still others form an operculum (a lid) at the top of the ovary, as in the Brazil nut family (*Lecythidaceae*).

Indehiscent fruits are derived from either single carpels or compound ovaries. Single carpel forms include the achene, the samara, and the caryopsis. Forms derived from a compound ovary include nuts and schizocarps. An achene is a fruit in which the single seed lies free in the cavity, attached only by a single point. The strawberry, for example, is really an aggregate fruit, and each "seed" is an achene. The samara is a winged achene and is found in the tree of heaven (*Ailanthus altissima;* Simaroubaceae) and ash (*Fraxinus;* Oleaceae). In the caryopsis, or grain, the seed adheres to the fruit wall (pericarp). The caryopsis is found among the cereal grasses, such as corn. Nuts have a stony pericarp, and usually only a single seed in each carpel matures, as in acorns of oaks (*Quercus;* Fagaceae) and hazelnuts (*Corylus avellana;* Betulaceae). Schizocarps are fruits in which each carpel of a compound ovary splits apart to form two or more parts, each with a single seed. Schizocarps are found in the carrot family (Apiaceae). Winged schizocarps are found in maples.

Seeds

Seeds are the mature ovules. They contain the developing embryo and the nutritive tissue for the seedling. Seeds are surrounded by one or two integuments, which develop into a seed coat that is usually hard. They are enclosed in the ovary of a carpel and thus are protected from the elements and predators.

The ovule is attached to the ovary wall until maturity by a short stalk called the funiculus. The area of attachment to the ovary wall is referred to as the placenta. The arrangement of placentae (placentation) in the compound ovary of angiosperms is characterized by the presence or absence of a central column in the ovary and by the site of attachment. In axile placentation the placentae are located on a central column; partitions from the central column to the ovary wall create chambers (locules) that separate the placentae and attached ovaries from each other. Free-central placentation resembles axile placentation; however, the column is not connected by partitions to the ovary wall, and thus no locules are formed. In basal placentation ovules are attached to the base of the ovary, and in parietal placentation the placentae are located directly on the ovary wall or on its extensions. Mature seeds are enclosed in integuments that may become hard and stony or that may have an outer fleshy, usually brightly coloured sarcotesta with an inner stony sclerotesta. Seed coats also may be winged or variously ornamented with prickles or sclerified hairs. In some seeds, there may be an extra covering, the aril, which is an outgrowth of the funiculus (e.g., the spice mace is derived from the red aril of *Myristica fragrans;* Myristicaceae). The aril of tomato seeds makes them slippery.

Mechanisms of Dispersal

Fruits and seeds are the primary means by which angiosperms are dispersed. The chief agents of dispersal are wind, water, and animals. Some fruits and seeds have modifications that aid in wind dispersal. Fruit modifications include samaras, samaroid schizocarps, and the feathery calyx lobes (e.g., dandelion). Seeds may be modified in various ways to promote dispersal: they may be extremely small and light (e.g., orchids, Orchidaceae), winged (e.g., common catalpa, *Catalpa bignonioides;* Bignoniaceae), plumed (e.g., milkweed), covered with woolly hairs (e.g., willows), or surrounded by explosive capsules that forcefully eject them into the air, as, for example, the touch-me-not (*Impatiens;* Balsaminaceae) and the witch hazel (*Hamamelis;* Hamamelidaceae). The fruits or seeds of many aquatic and shore plants are adapted to float on water as a means of dispersal; for this reason, coconuts (*Cocos nucifera;* Arecaceae) are readily transported across oceans to neighbouring islands. Adaptations for water dispersal include aerenchyma in fruits or seeds and light weight (e.g., water chestnut, *Trapa natans;* Lythraceae).

Animals disperse fruits and seeds either by ingesting and subsequently excreting them or by passively transporting them once they have adhered to an external part of the body, such as the fur or a claw. The evolution of fleshy fruits and seeds exemplifies the coevolution of plants and their animal agents of dispersal. An animal diet often consists solely of fruits and seeds that are designed to be eaten and dispersed, and in many cases these seeds require full or partial digestion to stimulate germination. Most fruits with a fleshy pericarp are eaten whole by vertebrates, including the stony endocarp or the stony seed coat. The seeds then either are regurgitated by the animal or pass through the alimentary canal and are excreted, often some distance from the original site. Seeds with an aril that encloses a stony seed coat or seeds with a sclerotesta and a fleshy, coloured sarcotesta are found in dehiscent fruits. They are eaten by animals after the fruit has ripened and split open. Often these seeds dangle from the fruit by long stalks (e.g., the follicles of *Magnolia*). The fleshy portion, whether originally a fruit or seed, is brightly coloured and sweet so as to attract vertebrates, particularly birds and mammals. Many fruits and seeds in the Amazon, however, are actually eaten and dispersed by fish during times of high water.

Inflorescences

Inflorescences are clusters of flowers on a branch or a system of branches. They are categorized generally on the basis of the timing of their flowering and by their arrangement on an axis. In indeterminate inflorescences, the youngest flowers, and therefore the last to open, are either at the top of the inflorescence (in elongated axes) or in the centre (in truncated axes). Branching and the associated flowers develop at some distance from the main stem (monopodial growth). Indeterminate inflorescences are of varied types: racemes, panicles, spikes, catkins (or aments), corymbs, and heads.

A raceme is an inflorescence in which a flower develops at the axil of each leaf along an elongated, unbranched axis. Each flower terminates a short stalk called a pedicel. The main axis has indeterminate growth; therefore, its growth does not cease at the onset of flowering. A spike is a raceme except that the flowers are attached directly to the axis at the axil of each leaf rather than to a pedicel. An example of a spike is the cattail (*Typha;* Typhaceae). The fleshy spike characteristic of the Araceae is called a spadix, and the underlying bract is known as a spathe.

A catkin (or ament) is a spike in which all the flowers are of only one sex, either staminate or carpellate. The catkin is usually pendulous, and the petals and sepals are reduced to aid in wind pollination when the inflorescence as a whole is shed. An example of a catkin is found in oaks. A corymb is a raceme in which the pedicels of the lower flowers are longer than those of the upper ones so that the appearance of the inflorescence overall is that of a flat flowerorganise. The lower flowers open first, and the axis of a corymb continues to produce flowers (indeterminate growth). Corymbs are found in the hawthorn (*Crataegus;* Rosaceae). If the axis is short or stunted, the flowers arise from a common point and appear to be at approximately the same level. This pattern, called an umbel, is actually a flattened raceme because the internodes of the axis, or peduncle (the point of origin of the leaves and flower axes), are shortened so that the pedicels are of the same length (e.g., the carrot family). A head is a raceme in which the peduncle is flattened and the flowers are attached directly to it (e.g., aster family, Asteraceae). This results in a grouping of small flowers in such a way as to appear as a single flower. In many members of the Asteraceae (e.g., sunflowers, *Helianthus annuus*), for instance, the outer (or ray) flowers have a well-developed zygomorphic corolla, and the inner (disk) flowers have a small actinomorphic corolla. The inner disk flowers generally are complete flowers, and the ray flowers generally are sterile.

In the compound indeterminate inflorescences, the main axis is branched so that the many inflorescences form off the main axis. A panicle is a branched raceme in which the branches are themselves racemes (e.g., yuccas, *Yucca*). In a compound umbel, all the umbel inflorescences arise from a common point and appear to be at about the same level (e.g., wild carrot). This organisation is the same for compound spikes, catkins, corymbs, and heads. The change from elongated axes (racemes and panicles) to flattened axes (corymbs and umbels) results in inflorescences in which the flowers are arranged close together. This close association encourages efficient pollination, and the extreme condensation of the inflorescences, as in the head, gives rise to an inflorescence that appears to be a single flower (e.g., sunflowers).

In the determinate (cymose) inflorescences, the youngest flowers (those that are the last to open) are at the bottom of an elongated axis or on the outside of a truncated axis (e.g., in the cymose umbel of onions, *Allium;* Alliaceae). These inflorescences are determinate because, at the time of flowering, the whole apical meristem produces a flower; thus, the entire axis ceases to grow. Each unit of a cyme consists of a dichasium, which has a central flower and two lateral flowers. The branching is primarily sympodial, and the inflorescence may be compound (e.g., catchfly, or campion, *Silene;* Caryophyllaceae). Many monocotyledons have a one-sided cyme called a helicoid cyme. A cymose inflorescence arranged in pairs at the nodes, in the manner of a false whorl, is called a verticillaster. Finally, there are mixed inflorescences, as, for instance, the cymose clusters arranged in a racemose manner (e.g., lilac, *Syringa vulgaris;* Oleaceae) or other types of combinations.

Reproduction

General Features

The vast array of angiosperm floral structures is for sexual reproduction. The angiosperm life cycle consists of a sporophyte phase and a gametophyte phase. The cells of a sporophyte body have a full complement of chromosomes (i.e., the cells are diploid, or $2n$); the sporophyte is the typical plant body that we see when we look at an angiosperm. The gametophyte arises when cells of the sporophyte, in preparation for reproduction, undergo meiotic division and produce reproductive cells that have only half the number of chromosomes (i.e., haploid, or n). A two-celled microgametophyte called a pollen grain germinates into a pollen tube and through division produces the haploid sperm. (The prefix *micro-* denotes gametophytes emanating from a male reproductive organ.) An eight-celled megagameto-phyte called the embryo sac produces the egg. (The prefix *mega-* denotes gametophytes emanating from female reprodu-ctive organs.)

Angiosperms are vascular plants, and all vascular plants have a life cycle in which the sporophyte phase (vegetative body) is the dominant phase and the gametophyte phase remains diminutive. In the nonvascular plants, such as the bryophytes, the gametophyte phase is dominant over the sporophyte phase. In bryophytes, the gametophyte produces its food by photosynthesis (is autotrophic) while the nongreen sporophyte is dependent on the food produced by the gametophyte. In nonseed vascular plants, such as ferns and horsetails, both the gametophyte and sporophyte are green and photosynthetic, and the gametophyte is small and without vascular tissue. In the seed plants (gymnosperms and angiosperms), the sporophyte is green and photosynthetic and the gametophyte depends on the sporophyte for nourishment. Within the seed plants, the gametophyte has become further reduced, with fewer cells comprising the gametophyte. The microgametophyte (pollen grain), therefore, is reduced from between 4 and 8 cells in the gymnosperms to a 3-celled microgametophyte in the angiosperms. A parallel reduction in the number of cells comprising a megagametophyte (ovule) has also taken place: from between 256 and several thousand cells in the gymnosperms to an 8-celled megagametophyte in most of the angiosperms. The significance of the reduction in megagametophyte cells appears to be related to pollination and fertilization. In many gymnosperms, pollination leads to the formation of a large gametophyte with copious amounts of stored starch for the nourishment of the potential embryo regardless of whether fertilization of the ovule can actually take place (i.e., whether the pollen is from the same species as the ovule). If the pollen is from a different species, fertilization or embryo development fails, so that the stored food is wasted. In angiosperms, however, the megagametophyte and egg are mature before the food is stored, and this is not ever accomplished until after the egg has been adequately fertilized and an embryo is present. This reduces the chances that the stored food will be wasted.

The process of sexual reproduction depends on pollination to bring these gametophytes in close association so that fertilization can take place. Pollination is the process by which pollen that has been produced in the anthers is received by the stigma of the ovary. Fertilization occurs with the fusion of a sperm with an egg to produce a zygote, which eventually develops into an embryo. After fertilization, the ovule develops into a seed, and the ovary develops into a fruit.

Anthers

A transverse section of the anther reveals four areas of tissue capable of producing spores. These tissues are composed of microsporocytes, which are diploid cells capable of undergoing meiosis to form a tetrad (four joined cells) of haploid microspores. The microspores become pollen grains and may eventually separate.

During pollen development, the layer of cells beneath the dermis of the anther wall (the endothecium) develops thickenings in the cell walls. The cell layer immediately inside the endothecium (the tapetum) develops into a layer of nutritive cells that either secrete their contents into the area around the microsporocytes or lose their inner cell walls, dissociate from each other, and become amoeboid among the microsporocytes. The pollen grains develop a thick wall of at least two layers, the intine and the exine. The intine, or inner layer, consists primarily of cellulose and pectins.

The exine, or outer layer, is composed of a highly decay-resistant chemical called sporopollenin. The exine usually has one or more thin areas, or pores, through which the pollen tubes germinate, and the thick area of the exine is usually highly sculptured. The number of pores and pattern of exine sculpturing are characteristic within an angiosperm family, genus, and often within a species.

The terminology to describe the various sculpturing patterns and position and number of pores is highly complex and only a basic description as related to functional aspects of sculpturing is given here. For example, smooth or essentially smooth pollen is loosely correlated with wind pollination, as in oaks (*Quercus*) and grasses (corn, *Zea mays*). Many plants pollinated by birds, insects, and small mammals have highly sculptured patterns of spines, hooks, or sticky threadlike projections by which pollen adheres to the body of the foraging pollinator as it travels to other flowers.

Each microspore (pollen grain) divides mitotically to form a two-celled microgametophyte; one cell is a tube cell (the cell that develops into a pollen tube), and the other is a generative cell, which will give rise to two sperm as a result of a further mitotic division. Thus, a mature microgametophyte consists of only three haploid cells—the tube cell and two sperm. Most angiosperms shed pollen at the two-celled stage, but in some advanced cases it is shed at the mature three-celled stage. When the pollen grains are mature, the anther wall either splits open (dehisces) longitudinally or opens by an apical pore.

Because the sporopollenin is resistant to decay, free pollen is well represented in the fossil record. The distinctive patterns of the exine are useful for identifying which species were present as well as suggesting the conditions of early climates. The proteins in the pollen walls are also a major factor in hay fever and other allergic reactions, and the spinose sculpturing patterns may cause physical irritation.

Ovules

An ovule is a saclike structure that produces the megaspores and is enclosed by layers of cells. This megasporangium is called the nucellus in angiosperms. After initiation of the carpel wall, one or two integuments arise near the base of the ovule primordium, grow in a rimlike fashion, and enclose the nucellus, leaving only a small opening called the micropyle at the top. In angiosperms the presence of two integuments is plesiomorphic (unspecialised), and one integument is apomorphic (derived). A single large megasporocyte arises within the nucellus near the micropyle and undergoes meiotic division, resulting in a single linear tetrad of megaspores. Three of the four megaspores degenerate, and the surviving one enlarges. The resulting megagametophyte produces the female gametes (eggs). This development (called megagametogenesis) involves free-nuclear mitotic divisions. The cell wall remains intact while the nucleus divides until the megagametophyte, or embryo sac, is formed. The embryo sac typically has eight nuclei. Free-nuclear mitotic division is also found in gametophyte formation in gymnosperms.

Four nuclei migrate to either end of the embryo sac. One nucleus from each group then migrates to the centre of the embryo; these become the polar nuclei. The two polar nuclei merge to form a fusion nucleus in the centre of the embryo sac. A cell wall develops around the fusion nucleus, leaving a central cell in the sac. Cell walls form around each of the chalazal nuclei to form three antipodal cells. During development, enlargement of the embryo sac leads to the destruction of most of the nucellus. This sequence of megasporogenesis and megagametogenesis, called the Polygonum type, occurs in 70 percent of the angiosperms in which the life cycle has been charted. Variations found in the remaining 30 percent represent derivations from the Polygonum type of seed development.

Pollination

Effective pollination involves the transfer of pollen from the anthers to a stigma of the same species and subsequent germination and growth of the pollen tube to the micropyle of the ovule. Pollen transfer is effected by wind, water, and animals, primarily insects and birds. Wind-pollinated flowers usually have an inconspicuous reduced perianth, long slender filaments and styles, covered with sticky trichomes and often branched stigmas, pendulous catkin inflorescences, and small, smooth pollen grains.

Wind pollination is derived in angiosperms and has developed independently in several different groups. For example, within the aster family wind pollination accompanied by floral reduction has developed independently in the tribes Heliantheae and Anthemideae. Water pollination occurs in only a few aquatic plants and is highly complicated and derived.

There is a wide range of animal pollinators of angiosperms as well as a wide range of adaptations by the flowers to attract those pollinators. Some of the living unspecialised families of basal angiosperms are pollinated by beetles. The beetles forage and feed on pieces of the perianth and stamens. There are no nectaries but rather food bodies on these organs.

Bees are responsible for the pollination of more flowers than any other animal group. Bees usually feed on nectar and in some cases on pollen. They may be general pollinators by visiting flowers of many species, or they may have adapted (i.e., elongated) their mouthparts to different flower depths and have become specialised to pollinate only a single species. Flowers pollinated by bees commonly have a zygomorphic, or bilaterally symmetrical, corolla with a lower lip providing a landing platform for the bee. Nectar is commonly produced either at the base of the corolla tube or in extensions of the corolla base. The bees partially enter the corolla mouth to feed with their long tongues on the nectar, at which point they deposit pollen picked up from other flowers and collect pollen from the new flower. Flowers pollinated by bees are often blue or yellow or exhibit patterns of both. Particular pattern markings and ultraviolet reflection patterns serve as recognition guides.

A high degree of coevolution is common in orchids (e.g., *Ophrys speculum* [see photograph]), where the flower not only appears to resemble the female wasp of a particular species but also produces the pheromone released by the insect to attract males of the species. The male wasp effects pollination by pseudocopulation with the orchid flower. Other insect pollinators include flies, butterflies, moths, and mosquitoes. Many flowers pollinated by flies are called carrion flowers because they look and smell like rotting meat. The skunk cabbage (*Symplocarpus foetidus*) and the carrion flowers (*Stapelia schinzii*) have evolved these characteristics independently.

Vertebrate pollinators include birds, bats, small marsupials, and small rodents. Many bird-pollinated flowers are bright red, especially those pollinated by hummingbirds. Hummingbirds rely solely on nectar as their food source. Flowers (e.g., *Fuchsia*) pollinated by birds produce copious quantities of nectar but little or no odour because birds have a very poor sense of smell. Flowers pollinated by bats produce large quantities of nectar and strong fragrances. They generally open only at night, when bats are the most active, and often hang down on long inflorescence stalks, which provide easy access to the nectaries and pollen. Some eucalypts (*Eucalyptus*) are pollinated by small marsupials (e.g., honey possums).

Whatever the agent of dispersal, the first phase of pollination is successful when a pollen grain lands on a receptive stigma. The surface of the stigma can be wet or dry and is often composed of specialised glandular tissue; the style is lined with secretory transmitting tissue. Their secretions provide an environment that nourishes the pollen tube as it elongates and grows down the style. If mitosis in the generative cell has not yet occurred in the pollen grain, it does so at this time.

To prevent self-fertilization, many angiosperms have developed a chemical system of self-incompatibility. The most common type is sporophytic self-incompatibility, in which the secretions of the stigmatic tissue or the transmitting tissue prevent the germination or growth of incompatible pollen. A second type, gametophytic self-incompatibility, involves the inability of the gametes from the same parent plant to fuse and form a zygote or, if the zygote forms, then it fails to develop. These systems force outcrossing and maintain a wide genetic diversity. The pollen tube ultimately enters an ovule through the micropyle and penetrates one of the sterile cells on either side of the egg (synergids). These synergids begin to degenerate immediately after pollination. Pollen tubes can reach great lengths, as in corn, where the corn silk consists of the styles for the corn ear and each silk thread contains many pollen tubes.

Fertilization and Embryogenesis

After penetrating the degenerated synergid, the pollen tube releases the two sperm into the embryo sac, where one fuses with the egg and forms a zygote and the other fuses with the two polar nuclei of the central cell and forms a triple fusion, or endosperm, nucleus. This is called double fertilization because the true fertilization (fusion of a sperm with an egg) is accompanied by another fusion process (that of a sperm with the polar nuclei) that resembles fertilization. Double fertilization of this type is unique to angiosperms. The zygote now has a full complement of chromosomes (i.e., it is diploid), and the endosperm nucleus has three chromosomes (triploid). The endosperm nucleus divides mitotically to form the endosperm of the seed, which is a food-storage tissue utilised by the developing embryo and the subsequent germinating seed. It has been shown that some of the most basal angiosperms actually form diploid endosperm, although they still experience double fertilization.

The three principal types of endosperm formation found in angiosperms—nuclear, cellular, and helobial—are classified on the basis of when the cell wall forms. In nuclear endosperm formation, repeated free-nuclear divisions take place; if a cell wall is formed, it will form after free-nuclear division. In cellular endosperm formation, cell-wall formation is coincident with nuclear divisions. In helobial endosperm formation, a cell wall is laid down between the first two nuclei, after which one half develops endosperm along the cellular pattern and the other half along the nuclear pattern. Helobial endosperm is most commonly found in the Alismatales (monocotyledons). In many plants, however, the endosperm degenerates, and food is stored by the embryo (e.g., peanut [groundnut], *Arachis hypogaea*), the remaining nucellus (known as perisperm; e.g., beet), or even the seed coat (mature integuments). Cellular endosperm is the least specialised type of endosperm with nuclear and helobial types derived from it.

The zygote undergoes a series of mitotic divisions to form a multicellular, undifferentiated embryo. At the micropylar end there develops a basal stalk or suspensor, which disappears after a very short time and has no obvious function in angiosperms. At the chalazal end (the region opposite the micropyle) is the embryo proper. Differentiation of the embryo—e.g., the development of cells and organs with specific functions—involves the development of a primary root apical meristem (or radicle) adjacent to the suspensor from which the root will develop and the development of one cotyledon (in monocotyledons) or two cotyledons (in dicotyledons) at the opposite end from the suspensor. A shoot apical meristem then differentiates between the two cotyledons or next to the single cotyledon and is the site of stem differentiation. The mature embryo is a miniature plant consisting of a short axis with one or two attached cotyledons. An epicotyl, which extends above the cotyledon(s), is composed of the shoot apex and leaf primordia; a hypocotyl, which is the transition zone between the shoot and root; and the radicle. Angiosperm seed development spans three distinct generations, plus a new entity: the parent sporophyte, the gametophyte, the new sporophyte, and the new innovation—namely, the endosperm.

Seedlings

Mature seeds of most angiosperms pass through a dormant period before eventually developing into a plant. The life span of angiosperm seeds varies from just a few days (e.g., sugar maple, *Acer saccharum*) to over a thousand years (e.g., sacred lotus, *Nelumbo nucifera*). Successful germination requires the right conditions of light, water, and temperature and usually begins with imbibition of water and the subsequent release from dormancy. During its early growth stages and before it has become totally independent of the food stored in the seed or cotyledons, the new plant is called a seedling.

Two patterns of seed germination occur in angiosperms, depending on whether the cotyledons emerge from the seed: hypogeal and epigeal. In hypogeous germination, the hypocotyl remains short and the cotyledons do not emerge from the seed but rather force the radicle and epicotyl axis to elongate out of the seed coat. The seed, with the enclosed cotyledons, remains underground, and the epicotyl grows up through the soil. When the cotyledons contain seed-storage products, these products are transferred directly to the developing radicle and epicotyl (e.g., garden pea). When the endosperm or perisperm contains the storage products, the cotyledons penetrate the storage tissues and transfer the storage products to the developing radicle and epicotyl (e.g., garlic, *Allium sativum*).

In epigeous germination, the radicle emerges from the seed and the hypocotyl elongates, raising the cotyledons, epicotyl, and remains of the seed coat above ground. The cotyledons may then expand and function photosynthetically as normal leaves (e.g., castor bean, *Ricinus communis*). When the cotyledons contain seed-storage products, they transfer them to the rest of the seedling and degenerate without becoming significantly photosynthetic (e.g., garden beans, *Phaseolus*). Eventually, the seedling becomes independent of the seed-storage products and grows into a mature plant capable of reproduction. Although the dispersal of seeds is essential in the reproduction and spread of angiosperm species, it is equally important for successful germination and seedling establishment to take place in an appropriate habitat.

Paleobotany and Evolution

The origins and diversity of flowering plants can best be understood by studying their fossil history. The fossil record provides important data to help show when and where early angiosperms lived, why flowering plants came to exist, and from what group or groups of plants they evolved. The earliest plants generally accepted to be angiospermous are known from the Early Cretaceous Period (about 145.5 million to 99.6 million years ago). Fossil pollen of angiosperms is first found in the Hauterivian and Barremian ages, which spanned from about 133.9 million to 125 million years ago. A very few angiosperm leaves and flowers are found in layers dating to the early Aptian Age (about 125 million to 112 million years ago). Many of the earliest fossils of angiosperms are most similar to small bushes or small herbaceous plants, such as those in the Chloranthaceae (Piperales), Ceratophyllaceae, and Ranunculaceae (Ranunculales) families. More diverse flora showing a larger variety of pollen, leaves, and reproductive organs with angiospermous affinities developed during the Albian Age (about 112 million to 99.6 million years ago).

From the end of the Albian (the close of the Early Cretaceous) and the beginning of the Late Cretaceous (about 99.6 million to 65.5 million years ago), angiosperms further diversified and dispersed. Many woody angiosperms evolved at that time, as did several modern groups, such as the magnolia, laurel, sycamore, and rose families. Herbaceous plants such as the water lilies (Nymphaeales), the family Ceratophyllaceae, and some of the early monocotyledons also persisted from the Albian until today. Because some of the oldest and most diverse angiosperm floras are found in Africa near the Equator, followed by low-latitude, angiosperm-dominated floras in North America, angiosperms are thought to have radiated from the Equator and spread to either pole. The angiosperms developed a close association with insect pollinators early in their evolution. This promoted outcrossing resulting in genetically vigorous offspring. Also, the relatively short generation time in which the angiosperms reproduce—permitting rapid population growth and easier colonization of disturbed habitats—gave the flowering plants an adaptive advantage over the gymnosperms, which were dominant during the Early Cretaceous. The seeds of angiosperms were small and were probably eaten and carried to new areas by animals.

Thus, the angiosperms were able to migrate into and occupy new areas of the world. At the beginning of the Cenomanian Age (about 99.6 million to 93.6 million years ago), angiosperms probably formed dominant pockets of vegetation along many low coastal tropical and warm temperate areas of the world. During the Cenomanian the angiosperms also spread to inland continental areas as well as northward and southward along the coasts. By the Middle to Late Cenomanian (about 95 million to 93.6 million years ago), angiosperms became the dominant form of vegetation in many areas of the world. One of the most conspicuous features of angiosperms is the flower. Most frequently, flowers are brightly coloured, often scented structures containing nectar and the male and female reproductive organs. Because it is important for the genetic integrity of a plant that it avoid pollinating itself or a nearby, possibly closely related, neighbour, pollen from one plant must be moved some distance to another plant. Wind is often an effective but imprecise pollination mechanism. Frequently, flowering plants are more accurately pollinated by animals, which carry the pollen some distance to another flower. Thus, development of showy flowers has involved the coevolution of insects or other animals and the early ancestors of the angiosperms.

Various groups of extinct seed plants have been proposed as the ancestral stock at different times in the evolution of the angiosperms. The Pteridospermales (seed ferns) are a group of extinct early seed plants that resemble small trees and shrubs with fernlike foliage. They bore seeds on their leaves or in specialised structures derived from leaves and had specialised pollen-bearing organs or simple anthers. The ovules and pollen organs were separate reproductive units, and wind may have been the most common agent of pollen transfer. Some seed ferns of the Paleozoic Era (about 542 million to 251 million years ago) contained pollen grains that were much too large to be effectively dispersed by the wind. These plants probably depended on insects to carry the pollen grains from one plant to another.

The Cycadeoidophyta are a group of extinct seed plants that contain members that have widely different reproductive structures. In some the female and male reproductive organs were separate, while in others the reproductive structures were organised into a common reproductive unit in which the male organs surrounded the female organ. These reproductive organs sat on a receptacle similar to that in flowering plants and often were surrounded by sterile bracts or leaflike tissue, which may have opened to form a flowerlike structure in the genus *Williamsoniella* (Cycadeoidales). Some extinct Cycadeoidales may have been pollinated by insects. The female and male reproductive organs tend to be clustered when insect pollination is involved, which is probably why most flowers are bisexual. It is not clear whether the flowering plants are derived from the Pteridospermales or the Cycadeoidales; however, in both groups the potential existed for modification of the plant body and the reproductive tissue to be responsive to both the physical and biological environments of the Mesozoic Era (about 251 million to 65.5 million years ago). The pollen evidence suggests that the Gnetales, a modern group of gymnosperms closely related to the angiosperms, were present during the Triassic Period (about 251 million to 199.6 million years ago). Thus, the evolution that produced the plants which were eventually recognised as the angiosperms must have been taking place during the Triassic, Jurassic, and earliest Cretaceous periods (which span from about 251 million to 99.6 million years ago).

The ancestral stock probably was a small to medium-size plant in which large leafy shoots contained individual fertile female, fertile male, and sterile leaves. The form of the plant was modified: the leaf size was reduced, and some shoots were modified so that the ovules remained enclosed inside the leaf tissue, which was shortened so that the ovule and pollen organs were borne close together. The sterile leaves may have been lost in some evolutionary lines or may have evolved into sepals and petals in others. The pollen-bearing organs (stamens) or ovule-bearing organs (carpels) may have been lost in some lines of evolution, resulting in unisexual flowers, or both may have been retained together in others to produce bisexual flowers.

Those early lines of angiosperm evolution in which wind may have functioned in pollination retained small, inconspicuous, often unisexual flowers. In those evolutionary lines that developed close associations with specific insect pollinators, the organs become dramatically modified. Small, inconspicuous bisexual or unisexual flowers are known from the Aptian Age. Large petals developed by the late Albian (about 105 million years ago). In insect-pollinated flowers and bisexual flowers that contain their characteristic nectaries, very large petals and anthers with abundant small pollen are known from the earliest Cenomanian Age. The presence of small, inconspicuous unisexual flowers, probably pollinated by wind or water, from the Aptian and late Albian suggests that the form and mode of reproduction of angiosperms were beginning to diverge from those of their ancestors even before this is attested in macrofossils. The special features of flowering plants that enhanced the coevolutionary links with animals evolved at various times in different groups of angiosperms. There were, however, three major nodes of coevolution in the development of flowering plants: the evolution of showy flowers attractive to animal (mainly insect) pollinators, the evolution of bilaterally symmetrical flowers with variously fused parts to direct the behaviour of particular animal pollinators (especially social insects and birds), and the evolution of larger energy-rich animals (especially mammals and birds) to disperse fruits and seeds. Each of these events had a dynamic effect on the evolution of angiosperms, increasing their diversity at different times in different groups and affecting their floral and fruit morphology in various ways.

The early angiosperms appear to have had few and radially arranged flower parts. The flowers were unisexual or bisexual, with superior ovaries, loosely closed to fully closed carpels, free flower parts, and small fruits and seeds. The fossil record of the early evolution of the flower demonstrates a tendency toward an increased number of flower parts, a loose to complete fusion of carpels, the development of a style, the elevation of the stigmatic surface upon the style, a slight increase in seed size, and a diversity of ways in which flowers were borne upon the plant.

The evolution of both female and male reproductive organs in the same flower was both beneficial and problematic in the early angiosperms. Insects visiting a unisexual flower either picked up pollen or deposited pollen, depending on the sex of the flower visited. Insect visits, therefore, only randomly fertilized flowers as the insect alternated between male and female flowers. It became beneficial to the flower to evolve a place for both sexes in a single flower so that each insect visit would deposit and remove pollen. When both sexes are present in a single flower, however, there develops a strong possibility that the flower may pollinate itself, a situation that would cause inbreeding depression, thereby reducing the vigour of the offspring over successive generations. It was probably very early in the evolutionary history of flowering plants that self-incompatibility was evolved, a mechanism that prevents flowers or plants from self-pollinating. The pollen of many modern insect-pollinated bisexual flowers is incompatible with the flower in which it is produced.

Another feature of flowers that developed as a result of insect pollination is pollen tube competition. When a pollen load of 50–200 pollen grains is deposited on a stigma at one time, each pollen grain grows a pollen tube into the stigmatic tissue. The pollen tubes that grow the fastest reach the ovules first and effect fertilization. It has been demonstrated that the pollen grain with the fastest-growing pollen tube carries genes that produce more vigorous offspring. By the Early Cenomanian the stigmas of some insect-pollinated flowers were elevated on styles, effectively establishing some distance for the pollen tubes to travel. This would establish pollen tube competition as a selective mechanism within some early flowers. During the first 70 million years of angiospermous evolution, all the known flowers were radially symmetrical. It is only in the early Paleogene Period—specifically, during the late Paleocene and early Eocene (about 58.7 million to 40.4 million years ago)—that the first evidence of bilaterally symmetrical flowers is found. The evolution of bilateral flowers—for example, that of the legumes and orchids—is an adaptation for specialised pollinators such as social insects (bees) and some birds. The sterile organs (sepals, petals) are modified to present a certain flower orientation to the pollinator, enabling the pollinator to enter the flower where the pollen organs and pollen-receptive tissue are positioned to maximize effective pollination. During the early Paleogene the bilateral organisation of floral organs coevolved with animal behaviour independently at different times and in various groups of angiosperms.

The evolution of mammals and birds also influenced the evolution of flowering plants in the early Paleogene. During the first 70–80 million years of their existence, the fruits and seeds of the angiosperms were small. The initial radiation of larger energy-rich fruits and seeds, such as the acorns, chestnuts, walnuts, legume pods, and the earliest grasses, took place during the Eocene. These fruits appeared over a short period of time contemporaneously with the diversification of seed- and fruit-eating mammals and birds. Seeds of fleshy fruits, such as grapes, also became common in the Eocene (about 45 million years ago). Thus, a second important node of plant and animal coevolution apparently developed about 50–60 million years ago, when angiosperms began to produce fruits and seeds that were attractive to animals. The animals served as agents to carry fruits and seeds some distance from the parent plant, further enhancing the potential for outcrossing and aiding in the dispersal of angiospermous plants to new areas of the world.

In summary, the evolutionary history of angiosperms is intimately but not exclusively tied to their coevolution with animal pollinators and agents of fruit and seed dispersal. Wind and water pollination and fruit and seed dispersal also continued throughout the entire evolutionary history of flowering plants. This network of evolutionary pressures resulted in the variety of flowers and fruits representative of present-day angiosperms. Accordingly, some of the most useful characters in determining the particular taxon to which living angiosperms belong are flowers, fruits, and seeds. The evolution of such vegetative characteristics as wood and leaves is more complex and less well understood.

Chapter 2

History of Plant Systematics

The history of plant systematics—the biological classification of plants—stretches from the work of ancient Greek to modern evolutionary biologists. As a field of science, plant systematics came into being only slowly, early plant lore usually being treated as part of the study of medicine. Later, classification and description was driven by natural history and natural theology. Until the advent of the theory of evolution, nearly all classification was based on the scala naturae. The professionalization of botany in the 18th and 19th century marked a shift toward more holistic classification methods, eventually based on evolutionary relationships.

Antiquity

Historians of botany trace the origins of botanical classification with folk taxonomy and ancient Greece. Theophrastus's (372–287 BC), a student of Aristotle, produced *Historia Plantarum*, the earliest surviving treatise on plants, where he listed the names of over 500 plant species. However he did not articulate a formal classification scheme; instead he relied on the common groupings of folklore combined with growth form: tree shrub; undershrub; or herb. The *Materia medica* of Dioscorides was also an important early compendium of plant descriptions (over five hundred); it was in use from its publication in the 1st century until the 16th century.

Early Modern Period

In the 16th century, works by Otto Brunfels, Hieronymus Bock, and Leonhart Fuchs helped to revive interest in natural history based on first-hand observation; Bock in particular included environmental and life cycle information in his descriptions. With the influx of exotic species in the Age of Exploration, the number of known species expanded rapidly, but most authors were far more interested in the medicinal properties of individual plants than an overarching classification system. Later influential Renaissance books include those of Caspar Bauhin and Andrea Cesalpino. Bauhin described over 6000 plants, which he arranged into 12 books and 72 sections based on a wide range of common characteristics. Cesalpino based his system on the structure of the organs of fructification, using the Aristotelian technique of logical division.

In the late 17th century, the most influential classification schemes were those of English botanist and natural theologian John Ray and French botanist Joseph Pitton de Tournefort. Ray, who listed over 18,000 plant species in his works, is credited with establishing the monocot/dicot division and some of his groups — mustards, mints, legumes and grasses — stand today (though under modern family names). Tournefort used an artificial system based on logical division which was widely adopted in France and elsewhere in Europe up until Linnaeus.

The book that had an enormous accelerating effect on the science of plant systematics was *Species Plantarum* (1753) by Linnaeus. It presented a complete list of the plant species then known to Europe, ordered for the purpose of easy identification using the number and arrangement of the male and female sexual organs of the plants. Of the groups in this book, the highest rank that continues to be used today is the genus. The consistent use of binomial nomenclature along with a complete listing of all plants provided a huge stimulus for the field.

Although meticulous, the classification of Linnaeus served merely as an identification manual; it was based on phenetics and did not regard evolutionary relationships among species. It assumed that plant species were given by God and that what remained for humans was to recognise them and use them (a Christian reformulation of the *scala naturae* or *Great Chain of Being*). Linnaeus was quite aware that the arrangement of species in the *Species Plantarum* was not a natural system, i.e. did not express relationships. However he did present some ideas of plant relationships elsewhere.

Modern and Contemporary Periods

Significant contributions to plant classification came from de Jussieu (inspired by the work of Adanson) in 1789 and the early nineteenth century saw the start of work by de Candolle, culminating in the *Prodromus*. A major influence on plant systematics was the theory of evolution (Charles Darwin published *Origin of Species* in 1859), resulting in the aim to group plants by their phylogenetic relationships. To this was added the interest in plant anatomy, aided by the use of the light microscope and the rise of chemistry, allowing the analysis of secondary metabolites.

Currently, the strict use of epithets in botany, although regulated by international codes, is considered unpractical and outdated. The very notion of species, the fundamental classification unit, is often up to subjective intuition and thus can not be well defined. As a result, estimate of the total number of existing "species" (ranging from 2 million to 100 million) becomes a matter of preference. While scientists have agreed for some time that a functional and objective classification system must reflect actual evolutionary processes and genetic relationships, the technological means for creating such a system did not exist until recently. In the 1990s DNA technology saw immense progress, resulting in unprecedented accumulation of DNA sequence data from various genes present in compartments of plant cells. In 1998 a ground-breaking classification of the angiosperms (the APG system) consolidated molecular phylogenetics (and especially cladistics or phylogenetic systematics) as the best available method. For the first time relatedness could be measured in real terms, namely similarity of the molecules comprising the genetic code.

Development of Plant Taxonomy

All cultures, whether based on hunting and gathering or advanced technology, place plants in groups and give names to those groups. Those cultures that are most directly connected to their raw plant materials tend to recognise such functional groups as food plants, poisonous plants, medicinal plants, plants that are useful for building shelters, and those that make good weapons. Within each of these groups, plants that look alike are apt to have similar names.

The Beginnings

A Greek, Theophrastus (370-285 BC) is the first person in the European tradition of plant taxonomy. He was concerned with identifying the essential features of plants - what aspect consistently differentiated one plant from another. To appreciate what an enormous question this was, and how great Theophrastus' contribution, read *The naming of names* by Anna Pavord (2005; Bloomsberry Press). Where Theophrastus stood out from his successors was his reliance on observation and scepticism concerning reports by others.

Theophrastus grouped plants on the basis of their habit and duration, recognising trees, shrubs, trees, annual, biennial, and perennial herbs. He must have been a keen observer because, in his book *Historia Plantarum,* he described 480 kinds of plant, recognising the difference between inflorescences in which flowering occurred from the outside inward as different from those in which it occurred in the reverse direction (i.e., centripetal versus centrifugal flowering), differences in ovary position, and in corollas with united petals versus those with separate petals. These seem minor items now, but to recognise them as somehow significant when everyone else simply says plant *a* is different from plant *b* is not easy. He did not appreciate how important flowers were to plants. He knew that shaking what we would now call male date flowers over female date flowers resulted in more fruit, but did not appreciate the significance of the pollen. Then again, how much did people really know about what happens during sex in animals at that time? If you have any interest in the history of ideas, read Pavord's book.

Albert Magnus appears to have been the first to notice the difference between monocots and dicots; he lived around 1200 A.D. Magnus even noticed differences in stem structure, a pretty incredible feat when you consider that your 10x hand lens is probably as good or better than any he had available. There was not a lot of progress in plant taxonomy during the next 400 years, particularly not in the grouping aspect. Knowing about plants fell in the realm of medicine, the chief interest of people being either in plants as food, medicine, poisons, or aphrodisiacs. There are relatively few plants that one can eat and these were soon known; fewer people knew which ones were medically important. Several herbals were published; they were often little more than copies of each other.

Post Printing

When printing became available in the early 1400s, the first botanical books printed were herbals. In these, the emphasis was on descriptions and accounts of the medicinal and domestic value of individual species. Some of these had a few beautiful illustrations; others had illustrations that owed a lot to the artist's imagination or faulty memory. All were enormously expensive by today's standards, but much cheaper than before the invention of printing.

Some herbalists contributed to the development of taxonomy as we understand it. For instance, Brunfels (1464-1534) distinguished between plants that had flowers and those that did not. Brock (1489-1554) (also known as Hieronymous) grouped the plants in his herbal according to their growth habit (i.e., as trees, shrubs, and herbs), but within each of these categories he "placed together, yet kept distinct, all plants which are related and connected, or otherwise resemble one another ". His herbal is also noteworthy for three other points. First, unlike most of his predecessors, Brock based his descriptions largely on his own observations; his predecessors relied heavily on descriptions of others, often going back as far as Theophrastus, Dioscorides, and Pliny -after all, *everyone* knew that the Greeks and Romans were the source of all knowledge. The other distinguishing features of Brock's herbal were 1) its incorporation of distribution information and 2) his debunking of much of the folklore concerning the supposed virtues or evils of particular plants.

At this time, it was not impossible for someone intrigued by plants to learn the names of all (or almost all) those with which he or she would come into contact, certainly all those that were in general use in Europe, probably around 1000 or so. (Just to give you some perspective on this: *Flora Europaea*, which was published in 1979, includes 11,577 species, many of which are introductions; the second edition of A *Utah Flora,* published in 1993, includes 3,284 species, 682 being introductions.

So long as Europe remained, in effect, isolated, there was no need for any systematic means of grouping or identifying plants; all one needed was a good memory. This was more or less the situation in 1500 A.D., but things were changing. It is convenient to regard publication of Linnaeus' *Species Plantarum* in 1753 as the end of the first stage in plant taxonomy or the beginning of the second but, before examining the reason for this, consider what was going on in the world between 1200 A.D. (Albert Magnus) and 1753 A.D. Five hundred and fifty years seems like a very long time to us, but bear in mind that change came more slowly before television, telephone, newspapers, trains, planes, and the internet.

Non-Botanical Developments

Marco Polo made his voyage to China in the thirteenth century (i.e., the 1200s). There had been trade with China prior to his trip, but the trade had been carried out through intermediaries; Europeans did not get far beyond the western Asia, i.e., Palestine (including what is now called Israel), the destination for the Crusaders. Consequently, Marco Polo's trip was truly amazing. He came back with tales of great wonders and great wealth. It is worth noting that the Arabs and other Muslims were more widely travelled than Europeans, who were most Christians, at that time. Many Muslims made the pilgrimage to Mecca where they would have come in contact with people from many different countries. Moreover, most travelled by land. Consquently, there were well known land routes to China and western Asian from North Africa.- whereas Europeans tended to think in terms of travelling by boat.

In 1453, Constantinople was captured by the Ottoman Empire (it had previously been held by the eastern branch of the Holy Roman Empire). The capture of Constantinople made anything originating in Asia or India very much more expensive for Europaeans because this meant almost the whole route to the east was under the control of the Ottoman Empire which demanded ever increasing tolls (charging what the traffic would bear). And there were strong memories of the Crusades that did not endear Europeans to the Otttomans, or vice versa. Sound familiar?

Prince Henry of Portugal pursued one possible answer to these higher prices: get to the East by going south from Portugal and around Africa. This took a lot of courage; it was unknown territory and there were lots of horror stories about what would happen to anyone that tried to do it. Nor was it obvious that one could go round Africa. The first Portuguese ship to sail past the equator did so in 1471. To the surprise of many, the sailors lived to tell the story. Fifteen years later, Bartholomew Diaz became the first European to round the Cape of Good Hope, but he did not go very far past it because his crew threatened to mutiny.

Not until 1498 was it clear that one could go round Africa to what, from Europe, is the Far East. In that year, Vasco da Gama rounded the Cape of Good Hope and landed in Calcutta, a city on the west coast of India. As a reward for opening up this trade route to the east, Vasco da Gama was awarded a large pension and the Portuguese title of 'Dom'. As some of you have probably realised, Columbus had discovered the West Indies six years earlier. Unfortunately for Columbus, his discovery was not financially rewarding; he died in poverty.

Once Columbus and da Gama had shown the way, several more expeditions went out, gradually exploring more of both the New and Old Worlds. Another landmark date is 1517. In that year, Luther nailed his 95 arguments against indulgences to the door of the cathedral in Wittenberg, challenging the pope. This was not the first time that the pope's authority or the teaching of the Roman Catholic church had been challenged in Europe, but this time the challenge stuck. People left the church and did not get killed by lightening bolts (although many were killed in the subsequent religions wars).

What has all this to do with plant taxonomy? First, although the impetus for the voyages of exploration was a mixture of commerce and religion, they also lead to more and more different plants (and animals) being brought to Europe or described (sometimes rather imaginatively) by those who went on the ships and it became harder and harder to for anyone to know all the different kinds of plants and animals. Some sort of system had to be developed 1) for placing the new plants in groups, whether new groups or old groups, and 2) for naming the groups. The first botanic gardens were founded in 1543 (Pisa) and 1545 (Padua) and around this time Ghini pioneered the idea of *hortus siccus* or a dried garden, what we would call a herbarium.

A factor favouring the development of such a system was that people were becoming bold enough to challenge old ways, old authorities. Knowledge of the classics, and ability to speak Latin and Greek were still essential for anyone wishing to be considered educated, but people were no longer so ready to believe a thing true simply because the church or a one of the classical scholars had said it.

Let us be honest here, 'people' in the preceding sentence refers to the small proportion of the population that was educated, but this proportion had grown enormously since the introduction of printing. In the 1500s it began to be expected that any well to do person should be able to read and write, not just priests and merchants. Moreover, it became more acceptable for intellectuals to devote their energy to consideration of the natural world rather than church doctrine.

Herbalists

The rapid increase in the number of plants known is reflected in the major herbals produced during this period. In 1583, Andrea Cesalpino, an Italian, wrote *De Plantis* (On Plants) in which he described about 1500 plants. He also outlined a basis for classifying, or grouping, plants, but it was not taken up. In 1650, Jean Bauhin, a French and Swiss physician, published the *Historia plantarum universalis* (Comprehensive Treatment of Plants) in which he treated 5000 species. Bauhin's work is also notable for 1) including the first good species descriptions, 2) distinguishing nomenclaturally between species and genera, and 3) using a binomial nomenclature. In 1703, a Brit, John Ray, published *Methodus plantarum* (Plant Systematics) in which he classified nearly 18,000 species according to their form and gross morphology. This represents a 10-fold increase in 120 years! Ray's groups are generally considered superior to Linnaeus', but there are good reasons why Ray's is considered less important than Linnaeus (as Linnaeus would have been the first to point out).

Linnaeus

So, what did Linnaeus do? Linnaeus provided an easy method of classifying (grouping) plants and, somewhat incidentally, a system of nomenclature. Linnaeus' grouping system was simple; all you needed to do to put a flowering plant in its correct group was count the number of different kinds of parts, and the number of stamens and pistils, and see how they were arranged [i.e., always in the same flower, sometimes in different flowers, always in different flowers, always in different plants, etc]. He even prepared what were, in effect, keys to his major groups.

After identifying the group to which a plant belonged, it was necessary to read each generic description through and see which one best matched your plant and then repeat this process to determine which species in the genus it belonged to. This sounds so simple, but no one had done it before, and no one did it for so many plants (and animals) as Linnaeus, as he himself pointed out. Another distinguishing feature of Linnaeus' work was the consistency of his descriptions, all of which were based on his own observations.

Linnaeus' system was an immense success because it made it possible for any botanist to place a new plant into the appropriate group relatively easily, whether or not he (or she, but I do not known of any women botanists from Linnaeus' time) had ever seen any other members of that group or not. It did not necessarily place closely related genera or species together; Linnaeus knew this, but that was not his aim. His aim was to provide an easy to use classification scheme, one that would enable everyone to sort things quickly and consistently. In this he succeeded, admirably.

Our present binomial system of botanical nomenclature also stems from Linnaeus, but it was not part of Linnaeus' aim. In his day, the name of a species species one was considered to be a generic name followed by a phrase that described which species of the genus one was referring to. (In Latin, adjectives usually follow nouns; Catholics are familiar with this structure in such phrases as "Mary -full of grace"). For instance, one plant was named "The Pontederia with cordate leaves and flowers in a spike". But this takes a long time to say and write so, *for* each species he described, Linnaeus put a single word in the margin, a different word *for* each species in a genus. Our scientific names are obtained by combining the generic name with this marginal name, The Pontederia with cordate leaves and flowers in a spike had the adjective "cordata" against it; we now refer to it as *Pontederia cordata*.

Linnaeus was not the first person to use binomials, but he was the first person to provide a binomial for every species he described, and he described a lot. Moreover, in his teaching, he routinely used the binomials to refer to the species, not the formal names. For this reason, Species Plantarum is taken as the starting point for botanical nomenclature. You will never see Bauhin or Ray as the authority for a scientific name. They used binomials before Linnaeus, but not so consistently nor for such a large number of plants.

Linnaeus' system was used for a long time in many parts of the world. One such botanist was Willdenow (1765-1812; his name appears as Willd. as an authority for many species), who edited the last edition of *Species plantarum* and was Director of the Berlin Botanical Gardens. Notable Americans employing the system were Schweinitz, Muhlenberg (Muhl.), Pursh, Walter, and Eaton. By 1820 or so, most taxonomists were no longer using it, preferring more 'natural' classifications, but even in the early twentieth century a guide to the Flora of India was published that used the Linnaean system, not because it was thought to reflect relationships but because it was a handy method for making a 'first sort' of poorly known plants. The analogy that comes to mind is cleaning up a room that is in a total mess. Sometimes it is best to put everything away somewhere first so that one has enough space to think through how one really wants to organise things in cupboards etc. Linnaeus gave us a first sort and some rules that enabled everyone else to do a first sort in the same manner. The next step was to develop a classification system that was more natural. Before leaving Linnaeus, he should be given credit *for* two other contributions. One was his ability to inspire others with his enthusiasm for getting to know as many different plants as possible. During his heyday, he had around 70 or so students in his classes.

The second contribution was setting the precedent for including a botanist on voyages of exploration. These botanists were his own students, most of whom initially sent their specimens back to him, but many went on to become well-known botanists in their own right. Peter Kalm was sent to America by the Universities of Abo and Uppsala. Hasselquist and Forskal sent specimens from Arabia, Syria, and Egypt.

Thunberg sent plants from the Cape of Good Hope and subsequently achieved a real coup, a post as surgeon in residence at a port in Japan. This made him the first European to collect extensively in Japan. In a very real way, Darwin's inclusion on the Beagle was part of the tradition started by Linnaeus.

French taxonomists were the only group that never used Linnaeus's system, sticking to that prepared by Tournefort, a Frenchman, until 1789, the year of the French Revolution and the year of publication of *Genera Plantarum* by the last in a family of French taxonomists, Antoine Laurent de Jussieu. To summarise: Linnaeus developed a logical system for assigning plants to groups, provided a major impetus for the universal use of a binomial system of nomenclature, inspired a large number of and students, persuaded those sponsoring exploring expeditions to include naturalists as an integral part of the expedition.

The next steps in taxonomy involved a) to improving on the groups recognised and b) formalizing the binomial system of nomenclature (naming). *Improving* on Linnaeus groups meant circumscribing [circumscribing = describing, with an emphasis on the boundaries of] more *natural groups.* Linnaeus himself recognised this need and published a book, *Systema Naturae,* in which he outlined such a system. His natural groups were not, however, as easy to recognise as his artificial groups, nor was it as easy to determine to which natural group a plant belonged.

Natural Systems

What is meant by a natural system? There has been quite a bit of argument about this, but basically it means that the groups recognised should reflect the relationships among plants; basically, this has, in the past, usually translated into placing plants with similar morphology [= shape, appearance] together. Remember, we are talking about the late 1700s and early 1800s. Darwin had not written his book on evolution and the mechanisms by which characteristics were passed on from one generation to another were unknown. It was recognised that offspring take features from both their parents and that hybrids are usually sterile, but thinking about the origin of species etc. was rather fuzzy. Biologists and geologists were becoming unhappy with the idea of a single creation event, but there was a lot to find out before the concept of descent by evolution could be articulated.

Among the discoveries made in the period 1753-1853 that contributed to a better understanding of plants were a) improvement in the understanding of optics and the consequent development of better microscopes, b) an awareness of the biological significance of the stamens and pistil in plants, and c) a better understanding of floral morphology. We shall come back to the topic of how to form taxonomic groups, but from a conceptual rather than historical point of view. First, for practical reasons, we need to distinguish between three much used (and abused) words in the context of plant taxonomy. These words are recognition, identification, and classification. Before going there, consider a couple of quotes:

Angiosperm Phylogeny Group

The Angiosperm Phylogeny Group, or APG, refers to an informal international group of systematic botanists who came together to try to establish a consensus on the taxonomy of flowering plants (angiosperms) that would reflect new knowledge about plant relationships discovered through phylogenetic studies. As of 2010, three incremental versions of a classification system have resulted from this collaboration published in 1998, 2003 and 2009. An important motivation for the group was, what they considered, deficiencies in prior angiosperm classifications since they were not based on monophyletic groups (i.e. groups that include all the descendants of a common ancestor).

APG publications are increasingly influential, with a number of major herbaria changing the arrangement of their collections to match the latest APG system.

Angiosperm Classification and the APG

In the past, classification systems were typically produced by an individual botanist or by a small group. The result was a large number of systems. Different systems and their updates were generally favoured in different countries. Examples are the Engler system in continental Europe, the Bentham & Hooker system in Britain (particularly influential because it was used by Kew), the Takhtajan system in the former Soviet Union and countries within its sphere of influence and the Cronquist system in the United States. Before the availability of genetic evidence, the classification of angiosperms (also known as *flowering plants*, *Angiospermae*, *Anthophyta* or *Magnoliophyta*) was based on their morphology (particularly of their flower) and biochemistry (the kinds of chemical compounds in the plant).

After the 1980s, detailed genetic evidence analysed by phylogenetic methods became available and while confirmed or clarified some relationships in existing classification systems, it radically changed others. This genetic evidence created a rapid increase in knowledge that led to many proposed changes; stability was "rudely shattered". This posed problems for all users of classification systems (including encyclopaedists).

In the late 1990s, an informal group of researchers from major institutions worldwide came together under the title of the 'Angiosperm Phylogeny Group' or APG. Their intention was to provide a widely accepted and more stable point of reference for angiosperm classification. Their first attempt at a new system was published in 1998 (the APG system). As of 2010, two revisions have been published, in 2003 (APG II) and in 2009 (APG III), each superseding the previous system. Eight researchers have been listed as authors to the three papers, and a further 33 as contributors.

A classification presents a view at a particular point in time, based on a particular state of research. Independent researchers, including members of the APG, continue to publish their own views on areas of angiosperm taxonomy. Classifications change, however inconvenient this is to users. However, the APG publications are increasingly regarded as an authoritative point of reference and the following are some examples of the influence of the APG system:

- A significant number of major herbaria, including Kew, are changing the order of their collections in accordance with APG.
- The influential World Checklist of Selected Plant Families (also from Kew) is being updated to the APG III system.
- In the USA, a recent photographic survey of the plants of the USA and Canada is organised according to the APG II system.

- In the UK, the latest edition of the standard flora of the British Isles (by Stace) is based on the APG III system. The previous edition was based on the Cronquist system.

Peter F. Stevens, one of the authors of all three of the APG papers, maintains a web site, APweb, hosted by the Missouri Botanical Garden, which has been regularly updated since 2001, and is a useful source for the latest research in angiosperm phylogeny which follows the APG approach.

Principles of the APG System

The principles of the APG's approach to classification were set out in the first paper of 1998, and have remained unchanged in subsequent revisions. Briefly, these are:

- The Linnean system of orders and families should be retained. "The family is central in flowering plant systematics." An ordinal classification of families is proposed as a "reference tool of broad utility". Orders are considered to be of particular value in teaching and in studying family relationships.
- Groups should be monophyletic (i.e. consist of *all* descendants of a common ancestor). The main reason why existing systems are rejected is because they do not have this property, they are not phylogenetic.
- A broad approach is taken to defining the limits of groups such as orders and families. Thus of orders, it is said that a limited number of larger orders will be more useful. Families containing only a single genus and orders containing only a single family are avoided where this is possible without violating the over-riding requirement for monophyly.
- Above or parallel to the level of orders and families, the term clades is used more freely. (Some clades have later been given formal names in a paper associated with the 2009 revision of the APG system.)

  The authors say that it is "not possible, nor is it desirable" to name all clades in a phylogenetic tree; however, systematists need to agree on names for some clades, particularly orders and families, to facilitate communication and discussion.

APG I (1998)

The initial 1998 paper by the APG made angiosperms the first large group of organisms to be systematically re-classified primarily on the basis of genetic characteristics. The paper explains the authors' view that there is a need for a classification system for angiosperms at the level of families, orders and above, but that existing classifications are "outdated". The main reason why existing systems are rejected is because they are not phylogenetic, i.e. are not based on strictly monophyletic groups (i.e. groups which consist of *all* descendants of a common ancestor). An ordinal classification of flowering plant families is proposed as a "reference tool of broad utility". The broad approach adopted to defining the limits of orders resulted in the recognition of 40 orders, compared to, for example, 232 in Takhtajan's 1997 classification.

Other features of the proposed classification include:

- Formal, scientific names are not used above the level of order, named clades being used instead. Thus eudicots and monocots are not given a formal rank on the grounds that "it is not yet clear at which level they should be recognised".
- A substantial number of taxa whose classification had traditionally been uncertain are given places, although there still remain 25 families of "uncertain position".
- Alternative classifications are provided for some groups, in which a number of families can either be regarded as separate or can be merged into a single larger family. For example, the Fumariaceae can either be treated as a separate family or as part of Papaveraceae.

A major outcome of the classification is the disappearance of the traditional division of the flowering plants into two groups, monocots and dicots. The monocots are recognised as a clade, but the dicots are not, with a number of former dicots being placed in separate groups basal to both monocots and the remaining dicots, the eudicots or 'true dicots'.

APG II (2003)

The second paper published by the APG presents an update to the original classification of 1998. The authors say that changes have been proposed only when there is "substantial new evidence" which supports them.

The proposed classification continues the tradition of seeking broad circumscriptions of taxa, for example trying to place small families containing only one genus in a larger group. The authors say that they have generally accepted the views of specialsts, although noting that specialists "nearly always favour splitting of groups" regarded as too varied in their morphology.

APG II continues and indeed extends the use of alternative 'bracketed' taxa allowing the choice of either a large family or a number of smaller ones. For example, the large Asparagaceae family includes 7 'bracketed' families which can either be considered as part of the Asparagaceae or as separate families.

Some of the main changes in APG II are:

- New orders are proposed, particularly to accommodate the 'basal clades' left as families in the first system.
- Many of the previously unplaced families are now located within the system.
- Several major families are re-structured.

In 2007, a paper was published giving a linear ordering of the families in APG II, suitable for ordering herbarium specimens, for example.

APG III (2009)

The third paper from the APG updates the system described in the 2003 paper. The broad outline of the system remains unchanged, but the number of previously unplaced families and genera is significantly reduced. This requires the recognition of both new orders and new families compared to the previous classification. The number of orders goes up from 45 to 59; only 10 families are not placed in an order and only two of these (Apodanthaceae and Cynomoriaceae) are left entirely outside the classification. The authors say that they have tried to leave long-recognised families unchanged, while merging families with few genera. They "hope the classification [...] will not need much further change." A major change is that the paper discontinues the use of 'bracketed' families in favour of larger, more inclusive families. As a result, the APG III system contains only 415 families, rather than the 457 of APG II. For example, the agave family (Agavaceae) and the hyacinth family (Hyacinthaceae) are no longer regarded as distinct from the broader asparagus family (Asparagaceae).

The authors say that alternative circumscriptions, as in APG I and II, are likely to cause confusion and that major herbaria which are re-arranging their collections in accordance with the APG approach have all agreed to use the more inclusive families.

In the same volume of the journal, two related papers were published. One gives a linear ordering of the families in APG III; as with the linear ordering published for APG II, this is intended for ordering herbarium specimens, for example. The other paper gives, for the first time, a classification of the families in APG III which uses formal taxonomic ranks; previously only informal clade names were used above the ordinal level.

Members of the APG

***Table:** Listed as "author" of one or more of the papers*

Name	*APG I*	*APG II*	*APG III*	*Institutional affiliation*
Birgitta Bremer	c	a	a	Swedish Academy of Sciences
Kåre Bremer	a	a	a	Uppsala University; Stockholm University
Mark W. Chase	a	a	a	Royal Botanic Gardens, Kew
Michael F. Fay	c	c	a	Royal Botanic Gardens, Kew
James L. Reveal		a	a	University of Maryland; Cornell University
Douglas E. Soltis	c	a	a	University of Florida
Pamela S. Soltis	c	a	a	Florida Museum of Natural History
Peter F. Stevens	a	a	a	Harvard University Herbaria; University of Missouri-St. Louis and Missouri Botanical Garden

a = listed as an author; c = listed as a contributor

Table: *Listed as "Contributor" to One or More of the Papers*

Name	*APG I*	*APG II*	*APG III*
Arne A. Anderberg	c	c	c
Anders Backlund	c		
Barbara G. Briggs	c		
Peter K. Endress	c		
Peter Goldblatt	c	c	
Mats H.G. Gustafson	c		
Sara B. Hoot	c		
Walter S. Judd	c	c	
Mari Källersjö	c	c	
Jesper Kårehed		c	
Elizabeth A. Kellogg	c		
Kathleen A. Kron	c	c	
Donald H. Les	c		
Johannes Lundberg		c	
Michael J. Moore			c
Cynthia M. Morton	c		
Daniel L. Nickrent	c	c	
Richard G. Olmstead	c	c	c
Bengt Oxelman		c	
J. Chris Pires		c	
Robert A. Price	c		
Christopher J. Quinn	c		
James E. Rodman	c	c	
Paula J. Rudall	c	c	c
Vincent Savolainen	c	c	
Kenneth J. Sytsma	c	c	c
David C. Tank			c
Mats Thulin	c		
Michelle van der Bank		c	
Kenneth Wurdack		c	c
Jenny Q.-Y. Xiang		c	c
Sue Zmarzty		c	c

c = listed as a contributor

Chapter 3

Angiosperm Taxonomy and Classification

For the past 264 years, since the publication of Linnaeus' *Systema Naturae,* botanists have been trying to improve the way in which plants are classified. French botanists, beginning with Adanson in 1763 and Jussieu in 1789, developed what they considered to be "natural systems", as opposed to that of Linnaeus' sexual system, which was admittedly artificial. The meaning of "natural system" continued to evolve, and with the addition of Darwin's concept of evolution it has come to mean something quite different than that of the French botanists. In the last 50 years, taxonomists have based classification on an evolutionary hierarchy, closely related species grouped together, first in genera, and then up to higher levels.

In this century, new categories of data have become available as the basis for classification. Morphology has long been the mainstay of taxonomy, but in the last 40 years new technologies have aided collection of anatomical and phytochemical data. The basic assumption however has remained that traits useful for identifying plants were also the best basis for classification. Recently it has been argued that all data should be integrated; it was not a matter of one category of information versus the others, but rather it was important to combine all data into one system. When DNA sequencing was developed in the mid-1980s, the stage was set for the inclusion of genetic studies (gene sequences). Computers were perfectly suited to these data, and the evaluation of relationships of all families of flowering plants became feasible as technology for sequencing DNA improved.

In 1993, Chase, Soltis, Olmstead and 39 other researchers published the first paper that evaluated the feasibility of analysing 500 DNA sequences from a representative sampling of families. There were several doubts about this process: could the computers and software then available deal with the complexity of such an enormous dataset, did gene sequences in spite of their functional constraints still contain evolutionary information, and would the patterns found with the first gene (the large subunit of RuBisCO, an important photosynthetic gene) be repeated with other genes? This first dataset took approximately 12 years to assemble, and no one was sure how long it would take to produce the next one, this time for a gene with a completely different product. It was clear that a gene with a completely different function should be used. About this time, automated sequencers were developed, allowing a rapid increase in gene sequencing. In 1997, this resulted in the publication by Soltis, Soltis, Nickrent et al. of the second large dataset for 18S nuclear ribosomal DNA, which produced similar patterns. Also in 1997, the results of a third large analysis of a gene, one involved in respiration in chloroplasts, were reported (by Savolainen, Chase, Hoot *et al.* and this too was highly similar. As important as the similarity of these studies was the fact that combining these data resulted in incredibly clear patterns of relationships. In 1998, the first large phylogenetic analysis of morphological, anatomical and phytochemical data was published by Nandi, Chase and Endress, and this too gave highly similar patterns.

All these analyses pointed in a single direction. All taxonomists, regardless of whether they worked on morphology or DNA, could now participate in the process of classification. More importantly, classification could now become predictive and refutable because it was based on data that were clearly visible to all researchers. If predictions failed, then the reasons should be sought and the classification improved. The new classification is thus both a tool for focusing research and a product of that research.

The new system of classification provides some wonderful examples of plants that previous taxonomists completely misunderstood, but it also demonstrates that large numbers of families were well understood. It is not a complete negation of all previous botanical research as some articles in the popular press have stated, neither does it overthrow the system of botanical nomenclature that began with Linnaeus nor refute Darwin. Quite the opposite, it draws upon a long history of debate and changing ideas about how the process of classification should be carried out. The new classification is constructed on the same lines of evidence and procedures that botanists have always used, but significantly in recent years this evidence has been augmented with massive amounts of genetic data, so that for the first time evolutionary patterns can be discerned clearly.

Some would assume that all these DNA data obviate the need for other kinds of study, but this is exactly the wrong conclusion. It makes the need for extension of all kinds of botanical study even more important because it is only through the improved knowledge of all types of data that predictivity can be enhanced. If classification were an end point in itself, then perhaps we could eliminate all other forms of botanical research except genetic studies, but classification is not an end at all. It is only the first step, and the really exciting prospects are those that we face now as we make use of the first predictive system of classification for any major group of organisms. Although the new system does not "turn the botanical world on its head" as one reporter recently wrote, it does represent a watershed in the way in which classifications are constructed. As another article in the popular press accurately reported, "a rose is still a rose, but everything else has indeed changed".

Angiosperm Families

Aceraceae

Aceraceae are a family of flowering plants also called the Maple Family. They contain two to four genera, depending upon the circumscription, of some 120 species of trees and shrubs. A common characteristic is that the leaves are opposite, and the fruit a schizocarp.

The maples have long been known to be closely related to the family Sapindaceae. Several taxonomists (including the Angiosperm Phylogeny Group) now include both the Aceraceae and the Hippocastanaceae in the Sapindaceae. Recent research (Harrington et al. 2005) has shown that while both Aceraceae and Hippocastanaceae are monophyletic in themselves, their removal from Sapindaceae *sensu lato* would leave Sapindaceae *sensu stricto* as a paraphyletic group, particularly with reference to the genus *Xanthoceras*.

Acorus

Acorus is a genus of monocot flowering plants. This genus was once placed within the family Araceae (aroids), but more recent classifications place it in its own family Acoraceae and order Acorales, of which it is the sole genus of the oldest surviving line of monocots. The exact relationship of Acorus to other monocots, howe-ver, is still debated by scientists. Some studies indicate that it is placed in a lineage (the order Alismatales), that also includes aroids (Araceae), Tofieldiaceae, and several families of aquatic monocots (e.g., Alismataceae, Posidoniaceae). Common names include Calamus and Sweet Flag. It is known as *vasambu* in Tamil language.

The name 'acorus' is derived from the Greek word 'acoron', a name used by Dioscorides, which in turn was derived from 'coreon', meaning 'pupil', because it was used in herbal medicine as a treatment for inflammation of the eye. The genus is native to North America and northern and eastern Asia, and naturalised in southern Asia and Europe from ancient cultivation. The known wild populations are diploid except for some tetraploids in eastern Asia, while the cultivated plants are sterile triploids, probably of hybrid origin between the diploid and tetraploid forms.

Taxonomy

Although the family Acoraceae was originally described in 1820, since then *Acorus* has traditionally been included in Araceae in most classification systems, as in the Cronquist system. The family has recently been resurrected as molecular systematic studies have shown that *Acorus* is not closely related to Araceae or any other monocot family, leading plant systematists to place the genus and family in its own order. This placement currently lacks support from traditional plant morphology studies, and some taxonomists still place it as a subfamily of Araceae, in the order Alismatales. The APG II system recognises order Acorales, distinct from the Alismatales, and as the sister group to all other monocots.

Characteristics

These plants are found in wetlands, particularly marshes, where they spread by means of thick rhizomes. Like many other marsh plants, they depend upon aerenchyma to transport oxygen to the rooting zone. They frequently occur on shorelines and floodplains where water levels fluctuate seasonally. The native North American species appears in many ecological studies. Relatively to other species of wetland plants, they have relatively high competitive ability. Although many marsh plants accumulate large banks of buried seeds, seed banks of *Acorus* may not accumulate in some wetlands owing to low seed production.

The seeds appear to be adapted to germinate in clearings; after a period of cold storage, the seeds will germinate after seven days of light with fluctuating temperature, and somewhat longer under constant temperature. A comparative study of its life history traits classified it as a "tussock interstitial", that is, a species that has a dense growth form and tends to occupy gaps in marsh vegetation, not unlike *Iris versicolour*. The inconspicuous flowers are arranged on a lateral spadix (a thickened, fleshy axis). Unlike aroids, there is no spathe (large bract, enclosing the spadix). The spadix is 4–10 cm long and is enclosed by the foliage. The bract can be ten times longer than the spadix. The leaves are linear with entire margin. The parallel-veined leaves of some species contain ethereal oils that give a sweet scent when dried. Fine-cut leaves used to be strewn across the floor in the Middle Ages, both for the scent, and for presumed efficacy against pests.

Regulation

Products derived from Acorus calamus were banned in 1968 as food additives by the United States Food and Drug Administration. The questionable chemical derived from the plant was â-asarone. Confusion exists whether all strains of Acorus Calamus contain this substance. Four varieties of Acorus calamus strains exist in nature; diploid, triploid, tetraploid and hexaploid. Diploids do not produce the carcinogenic â-asarone. Diploids are known to grow naturally in Eastern Asia (Mongolia and C Siberia) and North America. The triploid cytotype probably originated in the Himalayan region, as a hybrid between the diploid and tetraploid cytotypes. The North

American Calamus is known as Acorus Calamus var. Americanus or more recently as simply Acorus Americanus. Like the diploid strains of calamus in parts of the Himalayas, Mongolia, and C Siberia, the North American diploid strain does not contain the carcinogenic â-asarone. Research has consistently demonstrated that “â-asarone was not detectable in the North American spontaneous diploid Acorus [Calamus var. Americanus]”.

Species

In older literature and on many websites, there is still much confusion, with the name *Acorus calamus* equally but wrongfully applied to *Acorus americanus* (formerly Acorus calamus var. americanus).

The genus includes as many as six species:

- *Acorus americanus (Raf.) Raf.* (formerly known as *A. calamus* var. *americanus*) - American Sweet Flag; fertile diploid ($2n$ = 24); occurring in Alaska, Canada and northern USA. Diploid plants in Siberia and temperate Asia may also belong here, but have not been fully investigated (). Recently recognised as a distinct species by the *Flora of North America.*
- *Acorus calamus* L. - Common Sweet Flag; sterile triploid ($3n$ = 36); probably of cultivated origin. It is native to Europe, temperate India and the Himalayas and southern Asia, widely cultivated and naturalised elsewhere.
- *Acorus gramineus* Sol. ex Aiton - Japanese Sweet Flag or Grassy-leaved Sweet Flag; fertile diploid ($2n$ = 18); occurring in the Himalayas to Japan, Myanmar, Thailand, the Philippines.
- *Acorus triqueter* Turcz. ex Schott (syn. *A. calamus* var. *angustatus*) - fertile tetraploid ($4n$ = 48); occurring in eastern Asia, Japan and Taiwan.
- *Acorus latifolius* Z.Y.Zhu : native to China
- *Acorus xiangyeus* Z.Y.Zhu : native to China

Acorus from Europe, China and Japan have been planted in the United States.

Amborella

Amborella is a genus of rare understory shrubs or small trees endemic to the island of New Caledonia. The genus consists of only a single species, *Amborella trichopoda*, and is the only member of the family Amborellaceae. Wood of *Amborella* lacks the vessels characteristic of most flowering plants. It is of great interest to plant systematists because molecular phylogenetic analyses consistently place it at or near the base of the flowering plant lineage. That is, it represents a line of flowering plants that diverged very early on (about 130 million years ago) from all the other extant species of flowering plants. Comparing characteristics of this basal angiosperm, other flowering plants and fossils may provide clues about how flowers first appeared—what Darwin called the "abominable mystery".

Description

Amborella is a sprawling shrub or small tree up to eight metres high. It bears alternate or decussate, simple evergreen leaves without stipules. The leaves are two ranked, with distinctly serrated or rippled margins, and about 8 to 10 cm long. The species is dioecious: each flower produces both stamens and carpels, but only one sex develops fully and is fertile in the flowers of an individual plant. The small, creamy white, inconspicuous flowers are arranged in terminal panicles of 2 to 30 flowers borne in the axils of foliage leaves. Each flower is subtended by bracts and has a perianth of tepals (undifferentiated sepals and petals) arranged in a spiral or possibly whorled at the periphery. These features suggest that, as with other basal angiosperms, there is a high degree of developmental plasticity.

The bracts gradually transition into tepals, making it difficult to determine where a flower actually begins. Carpellate flowers are approximately 3 to 4 mm in diametre, with 7 or 8 tepals. There are 1 to 3 (or rarely none) well differentiated staminodes and a spiral of 4 to 8 free carpels (apocarpous). Carpels have green ovaries and lack a style. They contain a single ovule with the micropyle directed downwards. Staminate flowers are approximately 4 to 5 mm in diametre, with 6 to 15 tepals. These flowers bear 10 to 21 spirally arranged stamens, which become progressively smaller toward the centre. The innermost may be sterile. Filaments are represented by a short broad stalk. Anthers are triangular and consist of four pollen sacs, two on each side, with a small sterile central connective. Stamens have papillate secretory tips. A small pyramidal structure in the centre of the flower may represent a vestigial gynoecium. Typically, 1 to 3 carpels develop into fruit per flower. The fruit is an ovoid red drupe (approximately 5 to 7 mm long and 5 mm wide) borne on a short (1 to 2 mm) stalk. The remains of the stigma can be seen at the tip of the fruit. The skin is papery, surrounding a thin fleshy layer containing a red juice. The inner pericarp is lignified and surrounds the single seed. The embryo is small and surrounded by copious endosperm.

Phylogeny

This plant is currently accepted by plant systematists as the most basal lineage in the angiosperms clade. By "most basal", scientists mean that the Amborellaceae diverged the earliest from all other lineages of flowering plants. Comparing the derived characteristics that all other angiosperms share with each other, but not with the Amborella Family, may give scientists clues to what features early flowering plants had and how these characteristics have evolved through time. One early twentieth century idea of "primitive", or less derived, angiosperms that was accepted until relatively recently was modelled on the *Magnolia* blossom with numerous parts arranged in spirals on an elongated receptacle rather than the small numbers of parts in distinct whorls of more derived flowers. However, studies of a well-preserved fossil putative aquatic angiosperm, *Archaefructus*, have raised questions about what characteristics are more ancestral.

In a study designed to clarify relationships between the well-sequenced and well-studied model plants such as *Arabidopsis thaliana*, and the basal angiosperms such as *Amborella*, *Nuphar* of the Nymphaeaceae, *Illicium*, the monocots, and more derived angiosperms, the eudicots, scientists examined the chloroplast genomes and expressed sequence tags of these organisms, and other seed plants to create this cladogram. Note that in this image, the angiosperms are all of the plants not labelled "gymnosperms." This hypothesized relationship of the extant seed plants places *Amborella* as the sister taxon to all other angiosperms, and shows the gymnosperms as a monophyletic group sister to the angiosperms, supporting the theory that *Amborella* branched off earliest from all other living angiosperms. The dashed line between

Amborella and *Nuphar* is meant to indicate some uncertainty about the relationship between the Amborellaceae and the Nymphaeaceae, and whether or not they form a clade that is sister to the angiosperms, rather than *Amborella* alone being a monophyletic group sister to the angiosperms.

Because of its evolutionary position at the base of the flowering plant clade, there has been support for sequencing the complete genome of *Amborella trichopoda* to serve as a reference for evolutionary studies. In 2010, the US National Science Foundation began a genome sequencing effort in *Amborella*, and the draft genome sequence was posted on the project website in January 2012. The species was featured in the second episode of the BBC documentary series *How to Grow a Planet*, where Professor Iain Stewart described it as the "closest living relative of the first flower to evolve."

Classification

Amborella is placed alone in the family Amborellaceae. The APG II system recognised this family, but left it unplaced at order rank due to uncertainty about its relationship to the family Nymphaeaceae. In the most recent APG system, APG III, the Amborellaceae is placed in the monotypic order Amborellales at the base of the angiosperm phylogeny.

Ecology

This species is threatened in the wild by habitat destruction due to overgrazing, fire, mining (one of the most pervasive causes of ecosystem loss in New Caledonia), and urban expansion. *Amborella* is typically dioecious, but has been known to change sex in cultivation. *Amborella* has a mixed pollination system, relying both on insect pollinators and wind.

Apiaceae

The Apiaceae (or Umbelliferae), commonly known as carrot or parsley family, are a family of mostly aromatic plants with hollow stems. The family is large, with more than 3,700 species spread across 434 genera; it is the 16th-largest family of flowering plants. Included in this family are the well-known plants: angelica, anise, arracacha, asafoetida, caraway, carrot, celery, *Centella asiatica*, chervil, cicely, coriander (including cilantro), cumin, dill, fennel, hemlock, lovage, Queen Anne's lace, parsley, parsnip, sea holly, and the now extinct silphium.

Description

Most Apiaceae are annual, biennial or perennial herbs (frequently with the leaves aggregated toward the base), though a minority are shrubs or trees. Their leaves are of variable size and alternately arranged, or alternate with the upper leaves becoming nearly opposite. In some taxa, the texture is leathery, fleshy, or even rigid, but always with stomata. They are petiolate or perfoliate and more or less sheathing, the blade usually dissected and pinnatifid, but entire in some genera. Most commonly, crushing their leaves emits a marked smell, aromatic to foetid, but absent in some members. The flowers are nearly always aggregated in terminal umbels, simple or compound, often umbelliform cymes, rarely in heads.

The defining characteristic of this family is the inflorescence: a simple or compound umbel. Flowers across the Apiaceae are fairly uniform and are usually perfect (hermaphroditic) and actinomorphic, but some are andromonoecious, polygamomonoecious, or even dioecious (as in *Acronema*), with a distinct calyx and corolla, but the calyx if often highly reduced, to the point of being undetectable in many species, while the corolla can be white, yellow, pink or purple. The flowers are nearly perfectly pentamerous, with 5 petals, sepals, and stamens. The androecium contains of 5 stamens, but there is often variation in the functionality of the stamens even within a single inflorescence. Some flowers are functionally staminate (where a pistil may be present but has no ovules capable of being fertilized) while others are functionally pistillate (where stamens are present but their anthers do not produce viable pollen). Pollination of one flower by the pollen of a different flower of the same plant (geitonogamy) is common. The gynoecium consists of two carpels fused into a single, bicarpellate pistil with an inferior ovary.

When mature, the fused carpels separate into two mericarps. Stylopodiums secrete nectar, attracting pollinators like flies, mosquitoes, gnats, beetles, moths, and bees.

The fruits are nonfleshy schizocarp of two mericarps, each with a single seed; they separate at maturity and are dispersed by wind. Some fruit segments (like those in "Daucus" spp. are covered in bristles and spread via external transport. The seeds have an oily endosperm and generally contain large quantities of fatty oils, with the fatty acid petroselinic acid occurring universally throughout the family while rarely being found outside of the Apiaceae.

Systematics

The family is solidly placed within the Apiales order in the APG III classification system. It is closely related to Araliaceae and the boundaries between these families remain unclear. Traditionally groups within the family have been delimited largely based on fruit morphology, and the results from this have not been congruent with the more recent molecular phylogenetic analyses.

The subfamilial and tribal classification for the family is currently in a state of flux, with many of the groups being found to be grossly paraphyletic/polyphyletic.

Etymology

Apiaceae was first described by John Lindley in 1836. The name is derived from the type genus *Apium*, which was originally used by Pliny the Elder circa 50 AD for a celery-like plant. The alternative name for the family, Umbelliferae, derives from the inflorescence being generally in the form of a compound umbel. The family was one of the first to be recognised as a distinct group in Jacques Daleschamps' 1586 *Historia generalis plantarum*. With Robert Morison's 1672 *Plantarum umbelliferarum distribution nova* it became the first group of plants for which a systematic study was published.

Uses

Cultivation: Many members of this family are cultivated for various purposes. The plant structure includes a tap root, which can be large enough to be useful in food, as with parsnips (*Pastinaca sativa*), carrots (*Daucus carota*), and Hamburg parsley (*Petroselinum crispum*). Many plants of this group are also adapted to conditions that encourage heavy concentrations of essential oils, and as a result some are flavourful aromatic herbs. Examples are parsley (*Petroselinum crispum*), coriander (*Coriandrum sativum*), and dill (*Anethum graveolens*). The plentiful seeds of the umbels, likewise, are sometimes used in cuisine, as with, coriander (*Coriandrum sativum*), fennel (*Foeniculum vulgare*), cumin (*Cuminum cyminum*), and caraway (*Carum carvi*).

Other notable cultivated Apiaceae include chervil (*Anthriscus cerefolium*), angelica (*Angelica* spp.), celery (*Apium graveolens*), arracacha (*Arracacia xanthorrhiza*), poison hemlock (*Conium maculatum*), sea holly (*Eryngium* spp.), asafoetida (*Ferula asafoetida*), galbanum (*Ferula gummosa*), cicely (*Myrrhis odorata*), anise (*Pimpinella anisum*), lovage (*Levisticum officinale*), and hacquetia (*Hacquetia epipactis*).

Companion Plants

Almost every widely cultivated plant of this group is a companion plant. In large part, this is because the tiny flowers forming the umbels, are perfectly suited for ladybugs, parasitic wasps, and predatory flies, which actually drink nectar when not reproducing. They then will prey upon insect pests on nearby plants. Some of the plants are herbs that produce enough scent to possibly dilute the odours of nearby plants, or the pheromones emitted by insects that find those plants, which would otherwise attract more pests.

Folk Medicine

Many plants in this family, such as wild carrot, have estrogenic properties and have been used as a folk medicine for birth control. Most notable for this use is the extinct giant fennel, silphium. *Conium maculatum* has been used as a sedative and in treatments for arthritis and asthma in addition to its most famous use: as a "humane" method of killing criminals and philosophers. *Daucus carota* has been used as a diuretic, as a stimulant, to aid in uterine ailments, and one time as a treatment for jaundice. According to the Greek physician and botanist Dioscorides, "[Parsley] —especially Bastard or Wild Parsley—provokes venery and bodily lust and erection of the parts." Fennel has been used to treat nausea; species of Ferula have been used to ease constipation and coughs; cumin has been used as a treatment for diarrhea; coriander for halitosis; *Eryngium* spp. to treat genitourinary problems; Centella asiatica to treat syphilis; and *Ferula assa-foetida* has been used as a carminative, antispasmodic, and stimulant.

Other Uses

The poisonous members of the Apiaceae have been used for a variety of purposes globally. The poisonous *Oenanthe crocata* has been used to stupefy fish, *Cicuta douglasii* has been used as an aid in suicides, and arrow poisons have been made from various other family species.

Daucus carota has been used as colouring for butter and its roots used as a coffee substitute.

Dorema ammoniacum, *Ferula galbaniflua*, and *Ferula sumbul* are sources of incense.

The woody *Azorella compacta* Phil. has been used in South America for fuel.

Chemistry

Polyacetylenes can be found in Apiaceae vegetables like carrot, celery, fennel, parsley and parsnip where they show cytotoxic activities.

Apodanthaceae

The family Apodanthaceae comprises 22 to 30 species of endoparasitic herbs. They live in branches or roots of their hosts (as filaments similar to a fungal mycelium), emerging only to flower. The only leaves present are several bracts at the base of each flower. The plants do not carry out any photosynthesis (that is, they are holoparasitic). These are distributed into three genera: *Pilostyles*, *Apodanthes*, and *Berlinianche*. Mitochondrial and nuclear DNA sequences confidently place the Apodanthaceae in the Cucurbitales, where they also fit well in terms of their flower morphology.

Barbeya

Barbeya oleoides is the only species of its family (Barbeyaceae). It is a small tree native to the mountains of Somalia, Ethiopia, and the Arabian Peninsula. It can be found locally abundant in the transition zone between the dry, evergreen, Afromontane forests and lower-elevation evergreen bushlands. The Barbeyaceae family is closely related to its ecological associate on the Horn, the Dirachmaceae family. Evidence on the molecular level has demonstrated this despite obvious morphological differences between the two families such as Barbeya having small, unisexual, petalless flowers, while the flowers of Dirachmaceae are characterized by their bisexuality, and their relatively large petals (and size in general).

Bombacaceae

Bombacaceae were long recognised as a family of flowering plants or Angiospermae. The family name was based on the type genus *Bombax*. As is true for many botanical names, circumscription and status of the taxon has varied with taxonomic point of view, and currently the preference is to transfer most of the erstwhile family Bombacaceae to the subfamily Bombacoideae within the family Malvaceae in the order Malvales. The rest of the family were transferred to other taxa, notably the new family Durionaceae. Irrespective of current taxonomic status, many of the species originally included in the Bombacaceae are of considerable ecological, historical, horticultural, and economic importance, such as balsa, kapok, baobab and Durian.

Current Taxonomy

Recent phylogenetic research has shown that Bombacaceae as traditionally circumscribed (including tribe Durioneae) is not a monophyletic group. Furthermore, Bombacaceae is no longer recognised by the Angiosperm Phylogeny Group I 1998, II 2003 and Kubitzki system 2003 at the rank of family, the bulk of the taxa in question being treated as subfamily Bombacoideae within family Malvaceae *sensu lato*. A close relationship between Bombacaceae and Malvaceae has long been recognised but until recently the families have been kept separate in most classification systems, and continue to be separated in many references, including the reference work in classification of flowering plants: Heywood *et al.* 2007 and Takhtajan 2009, but have been lumped together in Angiosperm Phylogeny Website.

Heywood *et al.* say "although closely related to Malvaceae, molecular data supports their separation. Only pollen and habit seem to provide a morphological basis for the separation." On the other hand they say: "One approach is to lump them [the families in the core Malvales, including Bombacaceae] all into a 'super' Malvaceae, recognising them as subfamilies. The other, taken here, is to recognise each of these ten groups as families." As circumscribed in its traditional sense, the family

Bombacaceae includes around 30 genera (25 genera after Heywood *et al.*) with about 250 species of tropical trees, some of considerable girth, so called "bottle trees". Many species grow to become large trees, with *Ceiba pentandra* the tallest, reaching a height to 70 m. Several of the genera are commercially important, producing timber, edible fruit or useful fibres. The family is noted for some of the softest hardwoods commercially traded, especially Balsa, *Ochroma lagopus*. The fruit of the Durian, *Durio zibethinus* is famous, tasting better than it smells. At one time the fibre from the Kapok tree, *Ceiba pentandra* was used in making lifebuoys. The Baobabs or "Bottle trees" (*Adansonia* spp.) are important icons in certain parts of Africa, Australia and Madagascar, noted for their immensely stout trunk development, a mechanism for enhancing water storage.

Genera

- *Adansonia* L.
- *Aguiaria* Ducke
- *Bernoullia* Oliv.
- *Bombax* L.
- *Catostemma* Benth.
- *Cavanillesia* Ruiz & Pav.
- *Ceiba* Mill.
- *Chiranthodendron* Larreat. (according to Kubitzki in subf. Bombacoideae and considered more closely related to *Fremontodendron* by Baum *et al.* 2004)
- *Eriotheca* Schott & Endl.
- *Fremontodendron* Coville (according to Heywood *et al.*)
- *Gyranthera* Pittier
- *Huberodendron* Ducke
- *Matisia* Bonpl.
- *Neobuchia* Urb.
- *Ochroma* Sw.
- *Pachira* Aubl.
- *Patinoa* Cuatrec.
- *Pentaplaris* L.O.Williams & Standl. (according to Kubitzki in subf. Bombacoideae, but *incertae sedis*)

- *Phragmotheca* Cuatrec.
- *Pseudobombax* Dugand
- *Quararibea* Aubl.
- *Scleronema* Benth.
- *Septotheca* Ulbr.
- *Spirotheca* Ulbr. (according to Heywood *et al.*)

Genera of tribe Durioneae excluded from Bombacaceae after Heywood *et al.* 2007 and that should be included in Durionaceae

- *Boschia* Korth.
- *Coelostegia* Benth.
- *Cullenia* Wight
- *Durio* Adans.
- *Kostermansia* Soegeng
- *Neesia* Blume

Genus that should be excluded from Bombacaceae after Heywood *et al.* 2007 and that be included in Malvaceae s. s.

- *Camptostemon* Mast.

Genera considered synonym after Kubitzki 2003

- *Bombacopsis* Pittier = Pachira Aubl.
- *Chorisia* Kunth = Ceiba Mill.
- *Rhodognaphalon* (Ulbr.) Roberty = Pachira Aubl.

Genus not treated in Kubitzki

- *Lahia* Hassk., synonym of *Durio*, according to Mabberley

Brunellia

Brunellia is a genus of trees in the family Brunelliaceae. It consists of 62 species which grow in the mountainous regions from southern Mexico to Bolivia. There are 10 species in Ecuador, all but one of which occur in the Andean woodlands.

Classification

Brunellia contains the following species (but this list may be incomplete):

- *Brunellia acostae*, Cuatrec.
- *Brunellia almaguerensis*, Cuatrec.

- *Brunellia antioquensis*, (Cuatrec.) Cuatrec.
- *Brunellia boqueronensis*, Cuatrec.
- *Brunellia cayambensis*, Cuatrec.
- *Brunellia darienensis*, Cuatrec. & Porter
- *Brunellia ecuadoriensis*, Cuatrec.
- *Brunellia farallonensis*, Cuatrec.
- *Brunellia macrophylla*, Killip & Cuatrec.
- *Brunellia morii*, Cuatrec.
- *Brunellia occidentalis*, Cuatrec.
- *Brunellia ovalifolia*, Bonpl.
- *Brunellia pauciflora*, Cuatrec. & C.I.Orozco
- *Brunellia penderiscana*, Cuatrec.
- *Brunellia racemifera*, Tulasne
- *Brunellia rufa*, Killip & Cuatrec.
- *Brunellia subsessilis*, Killip & Cuatrec.
- *Brunellia zamorensis*, Steyerm.

Cabombaceae

Cabombaceae is the botanical name of a family of flowering plants. The family has been recognised as distinct by at least some taxonomists and by APG III (2009). The family consists of two genera of aquatic plants, (*Brasenia* and *Cabomba*), totalling half-a-dozen species. The APG system of 1998 included this family in the water lily family Nymphaeaceae, as did the APG II system, of 2003 (optionally). The family is part of the order Nymphaeales, which is one of the most basal flowering plant lineages.

Canna *(Plant)*

Canna (or canna lily, although not a true lily) is a genus of nineteen species of flowering plants. The closest living relations to cannas are the other plant families of the order Zingiberales, that is the Zingiberaceae (gingers), Musaceae (bananas), Marantaceae, Heliconiaceae, Strelitziaceae, etc.

Canna is the only genus in the family Cannaceae. Such a family has almost universally been recognised by taxonomists. The APG II system of 2003 (unchanged from the APG system,

1998) also recognises the family, and assigns it to the order Zingiberales in the clade commelinids, in the monocots. The species have large, attractive foliage and horticulturists have turned it into a large-flowered and bright garden plant. In addition, it is one of the world's richest starch sources, and is an agricultural plant.

Although a plant of the tropics, most cultivars have been developed in temperate climates and are easy to grow in most countries of the world as long as they can enjoy at least 6–8 hours average sunlight during the summer, and are moved to a warm location for the winter. See the *Canna* cultivar gallery for photographs of *Canna* cultivars. The name *Canna* originates from the Celtic word for a cane or reed.

Botany

The plants are large tropical and subtropical perennial herbs with a rhizomatous rootstock. The broad, flat, alternate leaves, that are such a feature of this plant, grow out of a stem in a long narrow roll and then unfurl. The leaves are typically solid green but some cultivars have glaucose, brownish, maroon, or even variegated leaves. The flowers are composed of three sepals and three petals that are seldom noticed by people, they are small and hidden under extravagant stamens. What appear to be petals are the highly modified stamens or staminodes. The staminodes number (1–) 3 (–4) (with at least one staminodal member called the labellum, always being present. A specialised staminode, the stamen, bears pollen from a half-anther. A somewhat narrower, 'petal' is the pistil which is connected down to a three-chambered ovary.

The flowers are typically red, orange, or yellow or any combination of those colours, and are aggregated in inflorescences that are spikes or panicles (thyrses). Although gardeners enjoy these odd flowers, nature really intended them to attract pollinators collecting nectar and pollen, such as bees, hummingbirds and bats. The pollination mechanism is conspicuously specialised. Pollen is shed on the style while still in the bud, and in the species and early hybrids some is also found on the stigma because of the high position of the anther, which means that they are self-pollinating. Later cultivars have a lower anther, and rely on pollinators alighting on the labellum and touching first the terminal stigma, and then the pollen. The wild species often grow to at least 2–3 m (6.6–9.8 ft) in height but there is a wide variation in size among cultivated plants; numerous cultivars have been selected for smaller stature.

Cannas grow from swollen underground stems, correctly known as rhizomes, which store starch, and this is the main attraction of the plant to agriculture, having the largest starch particles of all plant life.

Canna is the only member of the Liliopsida Class (monocot group) in which hibernation of seed is known to occur, due to its hard, impenetrable seed covering.

Distribution

The genus is native to tropical and subtropical regions of the New World, from the southern United States (southern South Carolina west to southern Texas) and south to northern Argentina. Although all cannas are native to the New World, they have followed mankind's journeys of discovery and some species are cultivated and naturalized in most tropical and sub-tropical regions.

Canna cultivars are grown in most countries, even those with territory above the Arctic Circle, which have short summers but long days, and the rapid growth rate of cannas makes them a feasible gardening plant, as long as they receive 6–8 hours of sunlight each day during the growing season and are protected from the cold of winter.

Origins

The first species of *Canna* introduced to Europe was *C. indica* L., which was imported from the East Indies, though the species originated from the Americas. Charles de l'Ecluse, who first described and sketched *C. indica* indicates this origin, and states that it was given the name of *indica*, not because the plant is from India, in Asia, but because this species was originally transported from America: "Quia ex America primum delata sit"; and at that time, one described the tropical areas of that part of the globe as the Western Indies.

Much later, in 1658, Pison made reference to another species which he documented under the vulgar or common name of 'Albara' and 'Pacivira', which resided, he said, in the *shaded and damp places, between the tropics*; this species is *Canna angustifolia* L., (later reclassified as *C. glauca* L. by taxonomists).

Without exception, all *Canna* species that have been introduced into Europe can be traced back to the Americas, and it can be asserted with confidence that *Canna* is solely an American genus. If Asia and Africa provided some of the early introductions, they were only varieties resulting from *C. indica* and *C. glauca* cultivars that have been grown for a long time in India and Africa, with both species imported from Central and South America.

Canna is an American genus, as pointed out by Lamarck where he argues that "Cannas were unknown to the ancients, and that it is only after the discovery of the New World, that they made their appearance in Europe". Since cannas have very hard and durable seed coverings, it is likely that seed remains would have survived in the right conditions and been found by archaeologists in the Old World. If the soils of India or Africa had produced some of them, they would have been imported before the 1860s into European gardens.

Uses

- Some species and many cultivars are widely grown in the garden in temperate and sub-tropical regions. Sometimes, they are also grown as potted plants. A large number of ornamental cultivars have been developed. They can be used in herbaceous borders, tropical plantings, and as a patio or decking plant.
- Internationally, cannas are one of the most popular garden plants and a large horticultural industry depends on the plant.
- The rhizome of cannas is rich in starch, and it has many uses in agriculture. All of the plant has commercial value, rhizomes for starch (consumption by humans and livestock), stems and foliage for animal fodder, young shoots as a vegetable and young seeds as an addition to tortillas.
- The seeds are used as beads in jewelry.

- The seeds are used as the mobile elements of the kayamb, a musical instrument from Réunion, as well as the hosho, a gourd rattle from Zimbabwe, where the seeds are known as *"hota"* seeds.
- In more remote regions of India, cannas are fermented to produce alcohol.
- The plant yields a fibre—from the stem—it is used as a jute substitute.
- A fibre obtained from the leaves is used for making paper. The leaves are harvested in late summer after the plant has flowered, they are scraped to remove the outer skin and are then soaked in water for 2 hours prior to cooking. The fibres are cooked for 24 hours with lye and then beaten in a blender. They make a light tan brown paper.
- A purple dye is obtained from the seed.
- Smoke from the burning leaves is said to be insecticidal.
- Cannas are used to extract many undesirable pollutants in a wetland environment as they have a high tolerance to contaminants.
- In Thailand, cannas are a traditional gift for Father's Day.
- In Vietnam, canna starch is used to make cellophane noodles known as *mi¿n dong*.

Taxonomy

Species: Although most cannas grown these days are cultivars, there are approximately 20 known species of the wild form, and in the last three decades of the 20th century, *Canna* species have been categorized by two different taxonomists, Paul Maas, from the Netherlands and Nobuyuki Tanaka from Japan. Both reduced the number of species from the 50-100 accepted previously, assigning most as synonyms.

This reduction in species is also confirmed by work done by Kress and Prince at the Smithsonian Institution, however, this only covers a subset of the species range.

See List of *Canna* species for full species information and descriptions.

Horticultural Varieties (Cultivars)

Cannas became very popular in Victorian times as garden plants and were grown widely in France, Germany, Hungary, India, Italy, the United Kingdom, and the USA. As tender perennials in northern climates, they suffered severe setbacks when two world wars sent the young gardening staff off to war. It took many years for the frugalities of war and its rationing subsequences to change to the more prosperous times of the late 20th century. The *Canna* genus has recently experienced a renewed interest and revival in popularity.

There were once hundreds of cultivars but many of these are now extinct. In 1910, Árpäd Mühle, from Hungary, published his *Canna* book, written in German. It contained descriptions of over 500 cultivars. In recent years many new cultivars have been created, but the genus suffers severely from having many synonyms for many popular ones. Most of the synonyms were created by old varieties re-surfacing without viable names, with the increase in popularity from the 1960s onwards. Research has accumulated over 2,800 *Canna* cultivar names, however, many of these are simply synonyms.

See List of *Canna* hybridists for details of the people and firms that created the current *Canna* legacy. In the early 20th century, Professor Liberty Hyde Bailey defined, in detail, two *garden species* (*C.* × *generalis* and *C.* × *orchiodes*) to categorise the floriferous cannas being grown at that time, namely the Crozy hybrids and the 'orchid-like' hybrids introduced by Carl Ludwig Sprenger in Italy and Luther Burbank in the USA, at about the same time (1894). The definition was based on the genotype, rather than the phenotype, of the two cultivar groups. Inevitably, over time those two floriferous groups were interbred, the distinctions became blurred and overlapped, and the Bailey species names became redundant. Pseudo-species names are now deprecated by the International Code of Nomenclature for Cultivated Plants which, instead, provides Cultivar Groups for categorising cultivars.

Agricultural Varieties

The Canna Agriculture Group contains all of the varieties of *Canna* grown in agriculture. "Canna achira" is a generic term used in South America to describe the cannas that have been selectively bred for agricultural purposes, normally derived from *C. discolour*. It is grown especially for its edible rootstock from which starch is obtained, but the leaves and young seed are also edible, and achira was once a staple food crop in Peru and Ecuador. Trials in Ecuador using a wide range of varieties have shown that achira can yield on average 56 tons of rhizomes and 7.8 of extractable starch per hectare. However the crop needs 9–12 months to mature to full productivity.

Many more traditional varieties exist worldwide, they have all involved human selection and so are classified as agricultural cultivars. Traditionally, *Canna "edulis"* has been reputed to be the variety grown for food in South America, but there is no scientific evidence to substantiate the name. It is probable that *C. edulis* is simply a synonym of *C. discolour*, which is also grown for agricultural purposes throughout Asia.

Cultivation

Cannas grow best in full sun with moderate water in well-drained rich or sandy soil. Cannas grow from perennial rhizomes but are frequently grown as annuals in temperate zones for an exotic or tropical look in the garden. In arid regions, cannas are often grown in the water garden, with the lower inch of pot submerged. In all areas, high winds tear the leaves so shelter is advised.

The rhizomes are frost tender and will rot if left unprotected in freezing conditions. In areas which go below about "10 °C (14 °F) in the winter (< USDA Zone 8), the rhizomes can be dug up before freezing and stored in a protected area (above 7 °C/ 45 °F) for replanting in the spring. Otherwise, it is recommended that cannas are protected by a thick layer of mulch overwinter.

The following is a list of canna cultivars that have gained the Royal Horticultural Society's Award of Garden Merit.

- 'Alaska'
- 'Amundsen'
- 'Alaska'
- 'Annaeei'
- 'Aphrodite'
- 'Delaware'

- *C.* × *ehemanii*
- 'Erebus'
- 'General Eisenhower'
- 'Ingeborg'
- 'Lenape'
- 'Louis Cayeux'
- 'Musifolia'
- 'Mystique'
- 'Phasion'
- 'Picasso'
- 'Roi Soleil'
- 'Russian Red'
- 'Schwäbische Heimat'
- 'Shenandoah'
- 'Verdi'
- 'Whithelm Pride'
- 'Wyoming'

Pests

Cannas are largely free of pests but in the USA plants sometimes fall victim to the Canna Leaf Roller and the resultant leaf damage, while not fatal to the plant, can be most distressing to a keen gardener. Slugs and snails are fond of cannas and can leave large holes in the leaves, preferring the tender young leaves that have not yet unfurled. Red spider mite can also be a problem for cannas grown indoors or during a very hot, long summer outdoors. Japanese beetles can also ravage the leaves if left uncontrolled.

Disease

Cannas are remarkably free of disease, compared to many genera. However, they may fall victim to canna rust, a fungus resulting in orange spots on the plant's leaves, caused by over moist soil. Cannas are also susceptible to certain plant viruses, some of which are *Canna* specific viruses, which may result in spotted or streaked leaves, in a mild form, but can finally result in stunted growth and twisted and distorted blooms and foliage.

The flowers are sometimes affected by a grey, fuzzy mold called botrytis. Under humid conditions it is often found growing on the older flowers. Treatment is to simply remove the old flowers, so the mold does not spread to the new flowers.

Propagation

Sexual Propagation

Seeds are produced from sexual reproduction, involving the transfer of pollen from the stamen of the pollen parent onto the stigma of the seed parent. In the case of *Canna*, the same plant can usually play the roles of both pollen and seed parents, technically referred to as a hermaphrodite. However, the cultivars of the Italian Group and triploids are almost always seed sterile, and their pollen has a low fertility level. Mutations are almost always totally sterile.

Canna seeds have a very hard seed coat, which contributes to their dormancy. Germination is facilitated by scarification of the seed coat, which can be accomplished by several techniques.

Pollination

The species are capable of self-pollination, but most cultivars require an outside pollinator. All cannas produce nectar and therefore attract nectar consuming insects, bats and hummingbirds that act as the transfer agent, spreading pollen between stamens and stigmas, on the same or different inflorescence.

Genetic Changes

Since genetic recombination has occurred a cultivar grown from seed will have different characteristics from its parent(s) and thus should never be given a parent's name. The wild species have evolved in the absence of other *Canna* genes and are deemed to be true to type when the parents are of the same species. In the latter case there is still a degree of variance, producing various varieties or minor forms (forma). In particular, the species *C. indica* is an aggregate species, having many different and extreme varieties and forms ranging from the giant to miniature, from large foliage to small foliage, both green and dark foliage and many different coloured blooms, red, orange, pink, and yellow and combinations of those colours.

Asexual Propagation

Division of Plant Parts: Outside of a laboratory, the only asexual propagation method that is effective is rhizome division. This is done by using material from a single parent, and as there is no exchange of genetic material such vegetative propagation methods almost always produce plants that are identical to the parent. After a summer's growth the horticultural cultivars can be separated into typically four or five separate smaller rhizomes, each with a growing nodal point ('growing eye'). Without the growing point, which is composed of meristem material, the rhizome will not grow.

Micropropagation

Micropropagation, or tissue culture as it is also known, is the practice of rapidly multiplying stock plant material to produce a large number of progeny plants. Micropropagation using in vitro methods that produce plants by taking small sections of plants and moving them into a sterile environment, where they first produce proliferations that are then separated from each other and then rooted or allowed to grow new stem tissue.

The process of plant growth is regulated by different ratios of plant growth regulators or PGRs, that promote cell growth. Many commercial organisations have attempted to produce cannas this way, and specifically the "Island Series" of cannas was introduced by means of mass-produced plants using this technique. However, cannas have a reputation of being difficult micropropagation specimens.

Micropropagation techniques can be employed on specimens infected with Canna virus and used to dis-infest plants of the virus, it is possible to use a growing shoot tip as the explant, the growing tip is induced into rapid growth, which results in rapid cell division that has not had time to be infected with the virus.

The rapidly growing region of meristem cells producing the shoot tip is cut off and placed in vitro, with a very high probability of being uncontaminated by virus, since it has not yet had contact with the sap of the plant which moves the virus within the plant. In this way, healthy stock can be reclaimed from virus contaminated plants.

Ceratophyllum

Ceratophyllum is a cosmopolitan genus of flowering plants, commonly found in ponds, marshes, and quiet streams in tropical and in temperate regions. They are usually called hornworts, although this name is also used for unrelated plants of the division Anthocerotophyta.

Ceratophyllum grows completely submerged, usually, though not always, floating on the surface, and does not tolerate drought. The plant stems can reach 1–3 m in length. At intervals along nodes of the stem they produce rings of bright green leaves, which are narrow and often much-branched. The forked leaves are brittle and stiff to the touch in some species, softer in others. The plants have no roots at all, but sometimes they develop modified leaves with a rootlike appearance, which anchor the plant to the bottom.

The flowers are small and inconspicuous, with the male and female flowers on the same plant. In ponds it forms thick buds (turions) in the autumn that sink to the bottom which give the impression that it has been killed by the frost but come spring these will grow back into the long stems slowly filling up the pond.

Hornwort plants float in great numbers just under the surface. They offer excellent protection to fish-spawn, but also to snails, infected with bilharzia. Because of their appearance and their high oxygen production, they are often used in freshwater aquaria.

Relationships and Classification

Ceratophyllum is considered unique enough to warrant its own family, Ceratophyllaceae, and its precise relationship to other angiosperms remains unclear. It was considered a relative of Nymphaeaceae and included in Nymphaeales in the Cronquist system but recent research has shown that it is not closely related to Nymphaeaceae or any other extant plant family. Some early molecular phylogenies suggested it was the sister group to all other angiosperms, but more recent ones have suggested that it is the sister group to either the monocots or the eudicots. The APG II system places the family in its own order, the Ceratophyllales.

The division of the genus into species is not completely settled. More than 30 species have been described, but many are probably just variants of these more widely accepted species:

- *Ceratophyllum demersum* L. (Rigid Hornwort or Common Hornwort)
- *Ceratophyllum echinatum* A.Gray (Spineless Hornwort)
- *Ceratophyllum muricatum* Cham. (Prickly Hornwort)
- *Ceratophyllum platyacanthum* Cham.
- *Ceratophyllum submersum* L. (Soft Hornwort or Tropical Hornwort)

Of these, *Ceratophyllum demersum* is widespread, with a global distribution; the others all have more restricted ranges.

Cochlospermaceae

Cochlospermaceae are a family of two genera and 20-25 species of trees and shrubs. They occur widely throughout the tropical regions of the world, but are curiously absent from Malaysia. Most species in this family are mesophytic or xerophytic, growing primarily in drier climates. Some authorities treat Cochlospermaceae as a synonym of Bixaceae. For example, the APG II system treats this as an optional segregate of Bixaceae.

Coriaria

Coriaria is the sole genus in the family Coriariaceae. It includes about 30 species of subshrubs, shrubs and small trees, with a widespread but disjunct distribution across warm temperate regions of the world, occurring as far apart as the Mediterranean region, southern and eastern Asia, New Zealand (where some are alpine species), the Pacific Ocean islands, and Central and South America.

The leaves are opposite or in whorls, simple, 2–9 cm long, without stipules. The flowers are borne in racemes 2–30 cm long, each flower small, greenish, with five small petals. The fruit is a small and shiny black (occasionally yellow or red) berry-like swollen corolla, highly poisonous in several species, though those of *C. terminalis* are edible. At least a few members of this genus are non-legume nitrogen fixers.

The Mediterranean species *C. myrtifolia* is known as Redoul, and the several New Zealand species are known by the Mâori name of tutu. The South American species *C. ruscifolia* is an evergreen climber known as Deu or Huique, and its fruits are used in Southern Chile to make rat poison.

Species List

An incomplete list of species of Coriaria:

- *Coriaria angustissima*
- *Coriaria arborea*
- *Coriaria japonica*
- *Coriaria kingiana*
- *Coriaria lurida*
- *Coriaria microphylla*
- *Coriaria myrtifolia*
- *Coriaria napalensis*
- *Coriaria plumosa*
- *Coriaria pottsiana*
- *Coriaria pteridoides*
- *Coriaria ruscifolia*
- *Coriaria sarmentosa*
- *Coriaria sinica*
- *Coriaria terminalis*
- *Coriaria thymifolia*

Dipentodon

Dipentodon is a genus of flowering plants in the family Dipentodontaceae. Its only species, Dipentodon sinicus, is a small, deciduous tree native to southern China, northern Myanmar, and northern India.[2] It has been little studied and until recently its affinities remained obscure.

Description

Dipentodon sinicus is a small, deciduous tree. The leaves are stipulate, alternate, and simple, with serrate margins. The inflorescence is variable in form, usually an abbreviated, umbelliform cyme containing 25 to 30 small flowers. The flowers are actinomorphic and yellowish green. The sepals and petals are only weakly differentiated, usually 5, rarely to 7 in number, free, or united only at the base. The hypanthium is very short or else the ovary is superior. The nectary disk is intrastaminal. The stamens are opposite the sepals. The ovary consists of three united carpels with two ovules per carpel. The ovary is 1-loculate, but partly 3-loculate at its base. The fruit is a 1-seeded drupaceous capsule.

History

Dipentodon was named and first described in 1911 by Stephen Troyte Dunn in what is now called the Kew Bulletin. At that time, Dunn wrote:

The name *Dipentodon*, proposed for it, refers to the most remarkable character possessed by the flowers in the exact similarity of the calyx teeth and petals (if I rightly call them so) and their insertion so nearly in one whorl that the appearance is given of a ten-toothed perianth.

Dipentodon was placed in its own family by Elmer Drew Merrill in 1941, but this placement was not generally followed. Instead, most authors put *Dipentodon* in the ill-defined and heterogeneous family Flacourtiaceae. In the twenty-first century, Flacourtiaceae is recognised by only a few taxonomists, and then only in a much narrower sense than it had been. *Dipentodon* is unrelated to Flacourtiaceae sensu stricto, a segregate of Salicaceae. Molecular phylogenetic studies have led to the widespread acceptance of the family Dipentodontaceae and its placement in the order Huerteales. Some authors have defined the family as consisting only of *Dipentodon*. Others, following the recommendation of a 2006 study, have included *Perrottetia*. When the APG II classification was published in 2003, the taxonomic position of *Dipentodon* was still unknown and it was placed incertae sedis in the angiosperms. It was listed in the appendix under *TAXA OF UNCERTAIN POSITION*.

Dirachma

Dirachma is the sole genus of the family Dirachmaceae. The genus has been monotypic until a second species was recently discovered in Somalia (Dirachma somalensis).

Eupomatia

Eupomatiaceae is the botanical name for a family of flowering plants. Such a family has been recognised by most taxonomists. The APG II system, of 2003 (unchanged from the APG system, of 1998), also recognises this family, and assigns it to the order Magnoliales in the clade magnoliids. The family consists of one genus only, *Eupomatia*, of three species, trees and shrubs, found on New Guinea and eastern Australia.

Flacourtiaceae

Flacourtiaceae is a defunct family of flowering plants whose former members have been scattered to various other families, mostly to Achariaceae, Samydaceae, and Salicaceae. It was so vaguely defined that hardly anything seemed out of place there and it became a dumping ground for odd and anomalous genera, gradually making the family even more heterogeneous. In 1975, Hermann Sleumer said

> *"Flacourtiaceae as a family is a fiction; only the tribes are homogeneous."*

In Cronquist's classification, it included 89 genera and more than 800 living species. Of these, many, including the type genus *Flacourtia*, have now been transferred to the Salicaceae in the molecular phylogeny-based classification, known as the APG II system, established by the Angiosperm Phylogeny Group. In the list below, Salicaceae is defined broadly. Some taxonomists would divide it into Salicaceae sensu stricto, Scyphostegiaceae, and Samydaceae.

Genera formerly included in Flacourtiaceae (current family in brackets)

- *Abatia* - (Salicaceae)
- *Aberia* - (Salicaceae)
- *Aphaerema* - (Salicaceae)
- *Azara* - (Salicaceae)
- *Banara* - (Salicaceae)
- *Bartholomaea* - (Salicaceae)
- *Barteria* - (Passifloraceae)
- *Bembicia* - (Salicaceae)

- *Bennettiodendron* - (Salicaceae)
- *Berberidopsis* - (Berberidopsidaceae)
- *Buchnerodendron* - (Achariaceae)
- *Byrsanthus* - (Salicaceae)
- *Calantica* - (Salicaceae)
- *Caloncoba* - (Achariaceae)
- *Camptostylus* - (Achariaceae)
- *Carrierea* - (Salicaceae)
- *Casearia* - (Salicaceae)
- *Chiangiodendron* - (Achariaceae)
- *Dioncophyllum* - (Dioncophyllaceae)
- *Dissomera* - (Salicaceae)
- *Dovyalis* - (Salicaceae)
- *Erythrospermum* - (Achariaceae)
- *Euceraea* - (Salicaceae)
- *Flacourtia* - (Salicaceae)
- *Gerrardina* - (Gerrardinaceae)
- *Gynocardia* - (Achariaceae)
- *Hasseltia* - (Salicaceae)
- *Hasseltiopsis* - (Salicaceae)
- *Hecatostemon* - (Salicaceae)
- *Hemiscolopia* - (Salicaceae)
- *Homalium* - (Salicaceae)
- *Hydnocarpus* - (Achariaceae)
- *Idesia* - (Salicaceae)
- *Itoa* - (Salicaceae)
- *Kiggelaria* - (Achariaceae)
- *Laetia* - (Salicaceae)
- *Lasiochlamys* - (Salicaceae)
- *Lindackeria* - (Achariaceae)
- *Ludia* - (Salicaceae)
- *Lunania* - (Salicaceae)
- *Macrohasseltia* - (Salicaceae)

- *Mocquerysia* - (Salicaceae)
- *Neopringlea* - (Salicaceae)
- *Neoptychocarpus* - (Salicaceae)
- *Olmediella* - (Salicaceae)
- *Oncoba* - (Salicaceae)
- *Ophiobotrys* - (Salicaceae)
- *Osmelia* - (Salicaceae)
- *Pangium* - (Achariaceae)
- *Phyllobotryon* - (Salicaceae)
- *Phylloclinium* - (Salicaceae)
- *Pineda* - (Salicaceae)
- *Pleuranthodendron* - (Salicaceae)
- *Poliothyrsis* - (Salicaceae)
- *Priamosia* - (Salicaceae)
- *Prockia* - (Salicaceae)
- *Pseudoscolopia* - (Salicaceae)
- *Pseudosmelia* - (Salicaceae)
- *Ryania* - (Salicaceae)
- *Ryparosa* - (Achariaceae)
- *Samyda* - (Salicaceae)
- *Scolopia* - (Salicaceae)
- *Tetrathylacium* - (Salicaceae)
- *Tisonia* - (Salicaceae)
- *Trichadenia* - (Achariaceae)
- *Trimeria* - (Salicaceae)
- *Xylosma* - (Salicaceae)
- *Xylotheca* - (Achariaceae)
- *Zuelania* - (Salicaceae)

Gerrardina

Gerrardina is a genus of two species of scrambling shrub found in eastern Africa. It is placed alone in family Gerrardiaceae. Placement are ordinal level is problematic; currently it is placed tentatively in Huerteales.

Gisekiaceae

The Gisekiaceae are a family of flowering plants. The family has not been recognised by most taxonomists. The APG II system (2003) does recognise the family and assigns it to the order Caryophyllales in the clade core eudicots. This represents a change from the APG system (1998), which did not recognise this family. The AP-Website assumes a placement of these plants in family Phytolaccaceae.

Gomortega

Gomortega keule (syn. *G. nitida*; Spanish names Keule, Queule and Hualhual) is a tree native to Chile. It is the sole species of the genus *Gomortega* and, according to the APG II system of 2003 (unchanged from the APG system of 1998), of the monotypic family Gomortegaceae, assigned to the order Laurales in the clade magnoliids. The plant grows only in a very narrow habitat range in coastal central Chile. The species is endangered of extinction due to overharvesting, clearing the forests where it is found for agriculture and silviculture.

Gomortega keule produces a yellow edible sweet fruit about 34–45 millimetres (1.3–1.8 in) in diametre, harvested for making a kind of marmalade.

Haptanthus

Haptanthus is the monotypic genus included in the flowering plants family Buxaceae, after molecular systematic research, and containing the sole species *Haptanthus hazlettii*, shrubs only known at the locality of Matarras, Arizona Municipality, Atlántida Department in Honduras.

Haptanthus has placed in its own family Haptanthaceae C. Nelson by Shipunov in 2005, but is considered among unplaced taxa (taxa *incertae sedis*) by APG (Stevens, P. F., Angiosperm Phylogeny Website, 2001 onwards) and others (including Goldberg) for insufficient data: fruits and seeds are still unknown. Haston *et al.* 2007 (also known as LAPG II) recognised this taxon (Haptanthaceae).

Goldberg (from Smithsonian Institution) has preferred not assign it to any family, after compare it with many other families, the most probable were Flacourtiaceae and Euphorbiaceae. Doust notes its resemblance to the Buxaceae and considers but does not accept various other families, in the end preferring to leave it unplaced. Attempts to sequence DNA from the one known sample failed.

Several attempts were made to recollect material from *Haptanthus*, but they succeeded only in 2010 with expedition of A. Shipunov and E. Shipunova. It is not clear whether *Haptanthus* is extinct; the area where it was originally found now mostly contains the introduced grass *Megathyrsus maximus*.

Illiciaceae

Illiciaceae is the botanical name of a family of flowering plants. The family has been recognised by most taxonomists, at least for the past several decades. The APG II system treated Illiciaceae as a family that may either be treated as part of another family, the Schisandraceae or allows for the optional segregation of the Illiciaceae from the Schisandraceae. The Illiciaceae as an optional segregate family then has the traditional circumscription of other taxonomic systems, consisting of a single genus, *Illicium*. The updated APG III system of 2009 does not recognise this family and includes *Illicium* in the Schisandraceae.

Description

The family consists of shrubs or small trees usually with volatile aromatic compounds. The leaves are evergreen, simple, alternate, spiral (sometimes crowded towards the tips of the twigs), leathery, petiolate, pinnately veined, non-sheathing, gland-dotted or not gland-dotted; also are aromatic, or without marked odour. The lamina is entire.

The flowers are hermaphrodite, and usually showy, ranging in size from small to large, and composed of numerous distinct parts that change grade slightly as they spiral around the receptacle. The parts begin with a perianth of whorled tepals, with the outer whorls more sepaloid, graduating to more petaloid inner tepals. The stamens are also numerous and in a whorl. The carpels are arranged in a whorl, are separate, and number from 5 to numerous carpels in each whorl, each carpel containing a single ovule. The stem presents nodes unilacunar (with one trace), with internal phloem absent, secondary thickening developing from a conventional cambial ring, xylem with tracheids; The sieve-tube plastids are S-type. The fruit is non-fleshy; the fruiting carpel is dehiscent, with a follicle (the cycle of follicles often spreading radially in a stellate pattern) and presents only one seed. The seeds are copiously endospermic and oily. The embryo is well differentiated (very small), achlorophyllous. The germination is phanerocotylar.

Phytochemistry

Leaves bear essential oils. Proanthocyanidins present (cyanidin and delphinidin). Flavonols present (kaempferol and quercetin). The fruits produce 2.5 – 5% of essential oil (Anetol, 80-90%, chavicol, safrol, shikimic acid); also contents veranisatins A, B and C and merrilactons.

Distribution and Species

The single genus, *Illicium*, has about 40 species native to sub-tropical and tropical Southeast Asia and the southeastern United States, the Caribbean, and parts of Mexico.

Species

- *Illicium verum*
- *Illicium floridanum*
- *Illicium henryi*
- *Illicium parviflorum*
- *Illicium anisatum*

For further details of placement in various systems see the entry for family Schisandraceae.

Ixerba

Ixerba brexioides, commonly known as Tawari, is an angiosperm tree. Until 2009 it was the single member of New Zealand's only endemic vascular plant family of Ixerbaceae. It is now placed with the New Caledonian genus *Strasburgeria* in family Strasburgeriaceae. It is only found in the North Island, between Waitomo and Kaitaia. It is used to make honey.

Memecylaceae

Memecylaceae are a family of flowering plants. The family includes about 430 species of trees and shrubs in seven genera. Memecylaceae are widespread in the tropics.

Misodendron

Misodendron is a genus of hemiparasites which grow as mistletoes on various species of *Nothofagus*. The twelve species are all restricted to South America. *Misodendron* is placed in its own family, Misodendraceae, in the order Santalales. The family has been recognised by many taxonomists, including the APG II system (2003; unchanged from the APG system of 1998), which assigns it to the Santalales in the clade core eudicots. These plants have the common name of feathery mistletoes.

Nelumbonaceae

Nelumbonaceae is the botanical name for a family of flowering plants. Such a family has been recognised by only a few taxonomists: the plants involved have often been included in family *Nymphaeaceae*. It is sometimes called the "lotus family" or "lotus-lily family".

The APG II system, of 2003 (unchanged from the APG system, of 1998), does recognise this family and places it in the order *Proteales*, in the clade eudicots.

The family consist of only a single genus, *Nelumbo*, of probably two species, of aquatic plants, found in North America and Asia (and perhaps some adjacent areas, but widely cultivated).

The Cronquist system, of 1981, also recognised this family but placed it in order *Nymphaeales* in subclass *Magnoliidae* in class *Magnoliopsida* [=dicotyledons].

The Dahlgren system and Thorne system (1992) also recognised this family and placed it in its own order *Nelumbonales* in superorder *Magnolianae* in subclass *Magnoliidae* [=dicotyledons].

Nymphaeaceae

Nymphaeaceae /ÌnjmfiÐÈejsiÐ/ is a family of flowering plants. Members of this family are commonly called water lilies and live in freshwater areas in temperate and tropical climates around the world. The family contains eight genera. There are about 70 species of water lilies around the world. The genus *Nymphaea* contains about 35 species across the Northern Hemisphere. The genus *Victoria* contains two species of giant water lilies and can be found in South America. Water lilies are rooted in soil in bodies of water, with leaves and flowers floating on the water surface. The leaves are round, with a radial notch in *Nymphaea* and *Nuphar*, but fully circular in *Victoria*.

Water lilies are divided into two main categories: hardy and tropical. Hardy water lilies bloom only during the day, but tropical water lilies can bloom either during the day or at night, and are the only group to contain blue-flowered plants.

Classification

Nymphaeaceae has been investigated systematically for decades because of the belief that they represent one of the earliest groups of angiosperms. Its position has been somewhat doubtful as the anatomy of these plants is more close to that of monocotyledons, while the venation of the leaves would indicate that they are dicotyledons. Nymphaeaceae is placed in the order Nymphaeales, which is the second diverging group of angiosperms after *Amborella* in the APG III-classification.

Nymphaeaceae is a small family of three to six genera: *Barclaya*, *Euryale*, *Nuphar*, *Nymphaea*, *Ondinea* and *Victoria*. The genus *Barclaya* is sometimes given rank as its own family, Barclayaceae, on the basis of an extended perianth tube (combined sepals and petals) arising from the top of the ovary and by stamens that are joined basally. However, molecular phylogenetic work includes it in Nymphaeaceae. The genus *Ondinea* has recently been shown to be a morphologically aberrant species of *Nymphaea*, and should be included in this genus. The genera *Euryale*, of far east Asia, and *VIctoria*, from South America, are closely related despite their geographic distance, but their relationship toward *Nymphaea* need further studies.

The sacred lotus was once thought to be a water-lily, but is now recognised to be a highly modified eudicot in its own family Nelumbonaceae of the order Proteales.

Description

The Nymphaeaceae are aquatic, rhizomatous herbs. The family is further characterized by scattered vascular bundles in the stems, and frequent presence of latex, usually with distinct, stellate-branched sclereids projecting into the air canals.

Hairs are simple, usually producing mucilage (slime). Leaves are alternate and spiral, opposite or occasionally whorled, simple, peltate or nearly so, entire to toothed or dissected, short to long Petiole (botanyiolate), with blade submerged, floating or emergent, with palmate to pinnate venation. Stipules are either present or absent.

Flowers are solitary, bisexual, radial, with a long pedicel and usually floating or raised above the surface of the water, with girdling vascular bundles in receptacle. Tepals are 4-12, distinct to connate, imbricate, and often petal-like. Petals lacks or 8 to numerous, inconspicuous to showy, often intergrading with stamens. Stamens are 3 to numerous, the innermost sometimes represented by staminodes.

Filaments are distinct, free or adnate to petaloid staminodes, slender and well differentiated from anthers to laminar and poorly differentiated from anthers; pollen grains usually monosulcate or lacking apertures. Carpels are 3 to numerous, distinct or connate. Fruit is an aggregate of nuts, a berry, or an irregularly dehiscent fleshy capsule. Seeds are often arillate, more or less lacking endosperm.

As Invasive Species

The beautiful nature of water lilies has led to their widespread use as ornamental plants. The Mexican water lily, native to the gulf coast of North America, is planted throughout the continent. It has escaped from cultivation and become invasive in some areas, such as California's San Joaquin Valley. It can infest slow moving bodies of water and is difficult to eradicate. Populations can be controlled by cutting top growth. Herbicides can also be used to control populations using glyphosate and fluridone.

Symbolism

The white water lily is the national flower of Bangladesh and State flower for Andhra Pradesh, India. The Blue waterlily is the national flower of Sri Lanka. It is also the birth flower for July. The water-lily has a special place in Tamil poetics, where it is considered symbolic of the grief of separation; it is considered to evoke imagery of the sunset, the seashore and the shark. Sangam landscape

Oncotheca

Oncotheca is a genus of tree endemic to New Caledonia. There are two species, *O. macrocarpa* and *O. balansae.*

Oncotheca is placed alone in family *Oncothecaceae.* This family has previously been placed in the Theales order, and in its own order, *Oncothecales,* but is currently left unplaced as to order in the asterids clade.

Pennantia

Pennantia is the sole genus in the plant family Pennantiaceae. In older classifications, it was placed in the family Icacinaceae. Most authorities have recognised three or four species, depending on whether they recognised *Pennantia baylisiana* as a separate species from *Pennantia endlicheri.* Mabberley, however, has recognised only two species. Pennantia is native to New Zealand, Norfolk Island, and northeast Australia.

Species

The following four species were recognised by Gardner and de Lange in 2002.

- *Pennantia baylisiana,* (W. Oliver) Baylis (Three Kings Islands)
- *Pennantia corymbosa,* 'kaikômako' (New Zealand)
- *Pennantia cunninghamii,* 'Brown Beech' (NSW and Queensland, Australia)
- *Pennantia endlicheri*

Schisandraceae

Schisandraceae is the botanical name of a family of flowering plants. Such a family has been recognised by most taxonomists, at least for the past several decades. Before that, the plants concerned were assigned to family Magnoliaceae.

The APG II system, of 2003, also recognises such a family. It places the family in order Austrobaileyales, which in turn is accepted as being among the most basic lineages in the clade angiosperms. APG II assumes this to be a family of three genera, Schisandraceae sensu lato. This family consists of woody plants, containing essential oils.

However, APG II does allow the option of segregating the genus *Illicium* as the family Illiciaceae. This leaves only two genera in this family, Schisandraceae sensu stricto. This consists of *Schisandra* and *Kadsura*, totalling several dozen species, found in tropical to temperate regions of East and Southeast Asia and the Caribbean.

The APG system, of 1998, recognised both the families Schisandraceae sensu stricto and Illiciaceae, unplaced as to order. It regarded both families as being among the most basic lineages in the clade angiosperms. The Cronquist system, of 1981, treated the plants in the family (in its wider sense) as two separate families, which together constituted

the order Illiciales,

in subclass Magnoliidae,

in class Magnoliopsida [=dicotyledons],

of division Magnoliophyta [=angiosperms].

Sonneratiaceae

Sonneratiaceae were a family of flowering plants placed in the order Myrtales by the Cronquist system. They consisted of two genera, *Sonneratia* and *Duabanga*. These are now generally placed in their own monotypic subfamilies of the family Lythraceae, making Sonneratiaceae superfluous. The family is named for French naturalist Pierre Sonnerat.

Sterculiaceae

Sterculiaceae is a botanical name for a group of flowering plants at the rank of family, which is now considered obsolete. As is true for any botanical name, the circumscription, status and placement of the taxon has varied with taxonomic point of view. The family name is based on the genus *Sterculia*.

As traditionally circumscribed the Sterculiaceae, Malvaceae, Bombacaceae, and Tiliaceae comprise the "core Malvales" of the Cronquist system and the close relationship among these families is generally recognised. Sterculiaceae may be separated from Malvaceae *sensu stricto* by the smooth surface of the pollen grains and the bilocular anthers.

Numerous phylogenetic studies have revealed that Sterculiaceae, Tiliaceae and Bombacaceae as traditionally defined are cladistically polyphyletic and the current status of each of these families is in doubt. The APG and APG II systems unite Bombacaceae, Malvaceae *sensu stricto*, Sterculiaceae and Tiliaceae into a more widely circumscribed Malvaceae, i.e., Malvaceae *sensu lato*.

In that view the taxa formerly classified in Sterculiaceae are treated in the subfamilies Byttnerioideae, Dombeyoideae, Helicteroideae and Sterculioideae of the Malvaceae *sensu lato*. The Thorne system takes an intermediate approach in combining the bulk of the traditional Sterculiaceae (but not including *Sterculia* itself) with elements of the traditional Tiliaceae to form the family Byttneriaceae.

Sterculiaceae had previously been recognised as a family by most systematists; in its traditional sense the family includes about 70 genera, totalling around 1,500 species of tropical trees and shrubs. The most famous products of the family are chocolate and cocoa from *Theobroma cacao*, followed by cola nuts. Many species yield timber.

Genera

- *Abroma* Jacq.
- *Acropogon* Schltr.
- *Aethiocarpa* Vollesen
- *Astiria* Lindl.
- *Ayenia* L.
- *Brachychiton* Schott & Endl.
- *Byttneria* Loefl.
- *Cheirolaena* Benth.
- *Chiranthodendron* Larreat.
- *Cola* Schott & Endl.
- *Commersonia* J.R.Forst. & G.Forst.
- *Cotylonychia* Stapf
- *Dicarpidium* F.Muell.
- *Dombeya* Cav.
- *Eriolaena* DC.
- *Firmiana* Marsili

- *Franciscodendron* B.Hyland & Steenis
- *Fremontodendron* Coville
- *Gilesia* F.Muell.
- *Glossostemon* Desf.
- *Guazuma* Mill.
- *Guichenotia* J.Gay
- *Hannafordia* F.Muell.
- *Harmsia* K.Schum.
- *Helicteres* L.
- *Helmiopsiella* Arenes
- *Helmiopsis* H.Perrier
- *Heritiera* Aiton
- *Hermannia* L.
- *Herrania* Goudot
- *Hildegardia* Schott & Endl.
- *Keraudrenia* J.Gay
- *Kleinhovia* L.
- *Lasiopetalum* Sm.
- *Leptonychia* Turcz.
- *Leptonychiopsis* Ridl.
- *Lysiosepalum* F.Muell.
- *Mansonia* J.R.Drumm. ex Prain
- *Maxwellia* Baill.
- *Megatritheca* Cristobal
- *Melhania* Forssk.
- *Melochia* L.
- *Neoregnellia* Urb.
- *Nesogordonia* Baill.
- *Octolobus* Welw.
- *Paradombeya* Stapf
- *Paramelhania* Arenes
- *Pentapetes* L.
- *Pimia* Seem.

- *Pterocymbium* R.Br.
- *Pterospermum* Schreb.
- *Pterygota* Schott & Endl.
- *Rayleya* Cristobal
- *Reevesia* Lindl.
- *Ruizia* Cav.
- *Rulingia* R.Br.
- *Scaphium* Schott & Endl.
- *Scaphopetalum* Mast.
- *Seringia* J.Gay
- *Sterculia* L.
- *Theobroma* L.
- *Thomasia* J.Gay
- *Triplochiton* K.Schum.
- *Trochetia* DC.
- *Trochetiopsis* Marais
- *Ungeria* Schott & Endl.
- *Waltheria* L.

Stixaceae

Stixaceae is a family in the plant order Brassicales. It is no longer recognised by most taxonomists. The three genera formerly included in Stixaceae — *Forchhammeria, Stixis* and *Tirania* — have sometimes been placed instead in the Capparaceae, but it is now clear that they do not belong there. It is unknown where they do belong though, so currently they are unplaced at family rank.

Chapter 4

Mycotoxicology—A Systems Approach

A 'system' may be viewed as a set of interacting components, where the interactions are just as important as the components themselves (after Open University, 1987). A 'systems' approach to the control of mycotoxins utilises conceptual models of interactions between, and within, commodity, spoilage, mycotoxin, and control subsystems. Within a system, the subsystems can freely interact; in other words, activity within one subsystem can influence events in one or more other subsystems. A better understanding of both the interactions and the components associated with these systems will assist in understanding the aetiology of mycotoxin production, and in formulating appropriate interventions for the control of mycotoxins and mycotoxicoses.

The Commodity System

Any commodity system is composed of numerous interacting technical and socioeconomic 'processes' including, for example, pest and disease control, harvesting, drying, processing, marketing, credit and pricing policies and cultural issues, to name but a few. A generalised, simplified commodity system is represented in Figure where selected processes are represented as interacting subsystems. At any point within the commodity system, the condition of the commodity is determined by a complex milieu involving a multitude of interactions between the crop, the macro-and micro-environment and a variety of biological, chemical, physical and socioeconomic factors. A change within any one process will invariably bring about changes in one or more of the other processes. Action taken before harvest to control pest damage and/or increase production can have a significant impact on the post-harvest quality of the commodity. Hybrid white maize, for example, has much higher yields than traditional varieties but has poor on-farm storage characteristics. Similarly, since it is very rare for a single commodity system to exist in isolation within a given agro-climatic region, it should be remembered that activities within one system can significantly effect events in other systems. Given the finite resources of farmers, for example, an increase in the importance of one commodity will frequently lead to the allocation of less resources towards the care of other commodities.

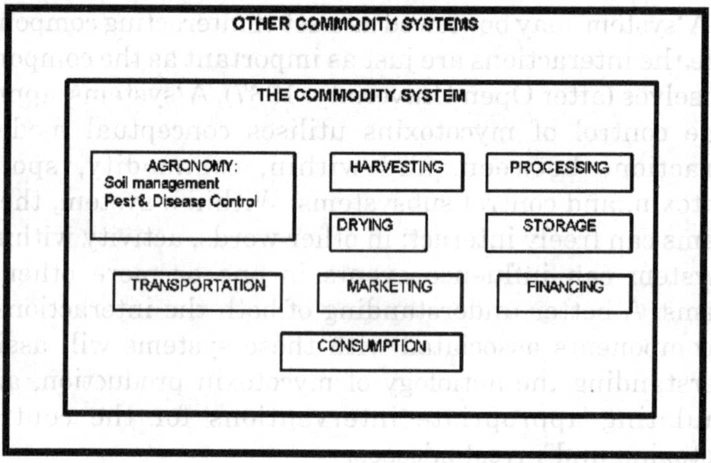

Figure: *The Commodity System*

The Spoilage System

Biodeterioration is the net result of numerous interacting spoilage agents which may be broadly described as biological, chemical, physical, macro-environmental and micro-environmental. However, the relative impact of these agents will often be largely determined by the nature and extent of human intervention. The factors which primarily contribute to bioterioration (including mould growth) within an ecosystem, are moisture, temperature and pests.

Moulds can grow over a wide range of temperatures and, in general, the rate of mould growth will decrease with decreasing temperature and available water. In grains, moulds utilise intergranular water vapour, the concentration of which is determined by the state of the equilibrium between free water within the grain (the grain moisture content) and water in the vapour phase immediately surrounding the granular particle.

The intergranular water concentration is described either in terms of the equilibrium relative humidity (ERH, %) or water activity (a_w). The latter describes the ratio of the vapour pressure of water in the grain to that of pure water at the same temperature and pressure, whilst the ERH is equivalent to the water activity expressed as a percentage.

For a given moisture content, different grains afford a variety of water activities and, consequently, support differing rates and type of mould growth. Typical water activities which are necessary for mould growth range from 0.70 to 0.99, the water activity, and the propensity for mould growth increasing with temperature. Maize, for example, can be relatively safely stored for one year at a moisture level of 15 per cent and a temperature of 15°C. However, the same maize stored at 30°C will be substantially damaged by moulds within three months.

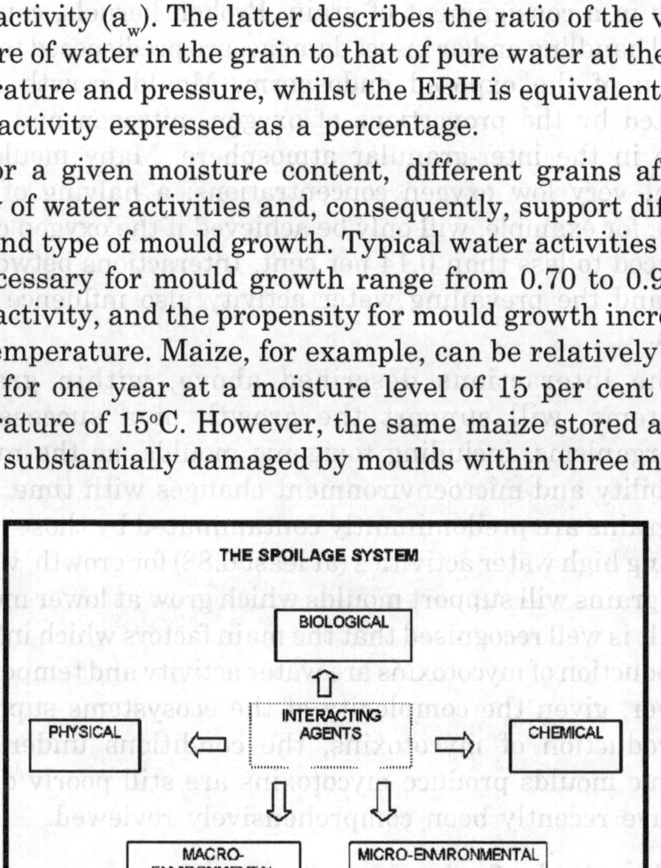

***Figure:** The Spoilage System*

Insects and mites (arthropods) can also make a significant contribution towards the biodeterioration of grain because of the physical damage and nutrient losses caused by their activity, and also because of their complex interaction with moulds and mycotoxins.

The metabolic activity of insects and mites causes an increase in both the moisture content and temperature of the infested grain. Arthropods also act as carriers of mould spores and their faecal material can be utilised as a food source by moulds. Furthermore, moulds can provide food for insects and mites but, in some case, may also act as pathogens. Another important factor that can affect mould growth is the proportion of broken kernels in a consignment of grain. Broken kernels, caused by general handling and/or insect damage, are predisposed to mould invasion of the exposed endosperm. Mould growth is also regulated by the proportions of oxygen, nitrogen and carbon dioxide in the inter-granular atmosphere. Many moulds will grow at very low oxygen concentrations; a halving of linear growth, for example, will only be achieved if the oxygen content is reduced to less than 0.14 per cent. Interactions between the gases and the prevailing water activity also influence mould growth.

The interactions described above, within granular ecosystems, will support the growth of a succession of microorganisms, including toxigenic moulds, as the nutrient availability and microenvironment changes with time. In the field, grains are predominantly contaminated by those moulds requiring high water activities (at least 0.88) for growth, whereas stored grains will support moulds which grow at lower moisture levels. It is well recognised that the main factors which influence the production of mycotoxins are water activity and temperature. However, given the complexity of the ecosystems supporting the production of mycotoxins, the conditions under which toxigenic moulds produce mycotoxins are still poorly defined; and have recently been comprehensively reviewed.

The Mycotoxin System

The Mycotoxin System may be considered in terms of three interacting subsystems: metabolism & toxicology; health & productivity; and wealth. After exposure (by ingestion, inhalation or skin contact), the toxicity of a mycotoxin is determined by a sequence of events (metabolism) involving the administration, absorption, transformation, pharmacokinetics, molecular interactions, distribution, and excretion of the toxin and its metabolites. In turn, the toxicity of a mycotoxin will be manifested by its effect on the health and productivity of crops, humans and animals; and, these effects will influence the production of wealth associated with human endeavour and agricultural and livestock products.

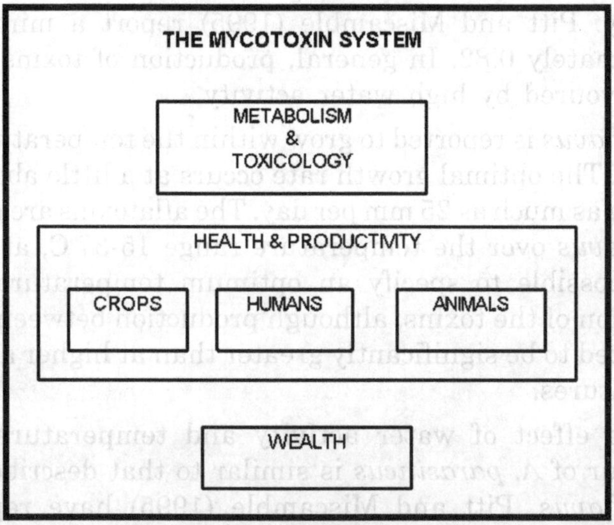

***Figure:** The Mycotoxin System*

Mycotoxins of worldwide Importance

Those moulds and mycotoxins which are currently considered to be of worldwide importance (Miller, 1994) are shown in Table. An 'important' mycotoxin will have demonstrated its capacity to have a significant impact upon human health and animal productivity in a variety of countries.

***Table :** Moulds and mycotoxins of worldwide importance*

Mould species	*Mycotoxins produced*
Aspergillus parasiticus	Aflatoxins B_1, B_2, G_1, G_2
Aspergillus flavus	Aflatoxins B_1, B_2
Fusarium sporotrichioides	T-2 toxin
Fusarium graminearum	Deoxynivalenol (or nivalenol) Zearalenone
Fusarium moniliforme *(F. verticillioides)*	Fumonisin B_1
Penicillium verrucosum	Ochratoxin A
Aspergillus ochraceus	Ochratoxin A

The Aflatoxins

The optimal water activity for growth of *A. flavus* is high (about 0.99). The maximum is at least 0.998 whereas the minimum water activity for growth has not been defined precisely. Pitt and Miscamble (1995) report a minimum of approximately 0.82. In general, production of toxins appears to be favoured by high water activity.

A. flavus is reported to grow within the temperature range 10-43°C. The optimal growth rate occurs at a little above 30°C, reaching as much as 25 mm per day. The aflatoxins are produced by *A. flavus* over the temperature range 15-37°C, at least. It is not possible to specify an optimum temperature for the production of the toxins, although production between 20-30°C is reported to be significantly greater than at higher and lower temperatures.

The effect of water activity and temperature on the behaviour of *A. parasiticus* is similar to that described above for *A. flavus*. Pitt and Miscamble (1995) have reported a minimum for growth of about 0.83; and a minimum for aflatoxin production of about 0.87. There are only limited data on the effect of temperature on the growth of *A. parasiticus* and the production of the aflatoxins. It was reported that optimal growth and toxin production occur at approximately 30 and 28°C, respectively. The term 'aflatoxins' was coined in the early 1960s when the death of thousands of turkeys ('Turkey X' disease), ducklings and other domestic animals was attributed to the presence of *A. flavus* toxins in groundnut meal imported from South America.

The aflatoxin-producing moulds occur widely, in temperate, subtropical and tropical climates, throughout the world; and the aflatoxins may be produced, both before and after harvest, on many foods and feeds especially oilseeds, edible nuts and cereals. Although the aflatoxins are predominantly associated with commodities of subtropical and tropical origin, their occurrence has also been reported in temperate climates in acid-treated grains.

Aflatoxin B_1 is a human carcinogen and is one of the most potent hepatocarcinogens known. Human fatalities have also occurred from acute aflatoxin poisoning in India, for example, when unseasonal rains and a scarcity of food prompted the consumption of heavily contaminated maize. If the immunosuppressive action of the aflatoxins in livestock is similarly manifested in humans, it is possible that the aflatoxins (and other mycotoxins) could play a significant role in the aetiology of human disease in some developing countries, where a high exposure to these toxins has been reported.

Lubulwa and Davis (1994) have studied economic losses attributable to the occurrence of aflatoxin only, in maize and groundnuts, in Southeast Asian countries (Thailand, Indonesia and the Philippines).

They concluded that contaminated maize accounted for about 66 per cent of the total loss, whereas losses attributable to spoilage and deleterious effects on human and animal health were 24, 60 and 16 per cent of the total, respectively. However, the study considered losses associated with morbidity and premature death caused by cancer only. Consequently, it is likely that when the additional effects on human health caused by the immunotoxic effect of aflatoxin (and other mycotoxins) are included, the loss associated with aflatoxins will be significantly increased.

The Trichothecenes

Surprisingly little is known about the effects of water activity and temperature on the behaviour of the *Fusarium* moulds, including the production of mycotoxins.

In the case of *F. graminearum*, the temperature limits for growth have not been reported, although the optimal temperature has been estimated at 24-26 °C. The minimum water activity for growth is 0.9; the maximum limit is recorded as in excess of 0.99. No information is available on the effect of water activity and temperature on the production of deoxynivalenol, nivalenol and zearalenone.

The minimum water activity for the growth of *F. sporotrichioides* is 0.88, whereas the maximum limit is reported as >0.99. The minimum, optimal and maximum temperatures for growth are-2.0, 22.5-27.5 and 35°C, respectively. As with the other *Fusarium* moulds, there is no information on the conditions required for the production of T-2 toxin.

T-2 toxin and deoxynivalenol belong to a large group of structurally-related sesquiterpenes known as the ´trichothecenes'.

T-2 toxin is produced on cereals in many parts of the world and is particularly associated with prolonged wet weather at harvest. It is the probable cause of ´alimentary toxic aleukia' (ATA), a disease which affected thousands of people in Siberia during the Second World War, leading to the elimination of entire villages.

The symptoms of ATA included fever, vomiting, acute inflammation of the alimentary tract and a variety of blood abnormalities. T-2 toxin is responsible for outbreaks of haemorrhagic disease in animals and is associated with the formation of oral lesions and neurotoxic effects in poultry.

The most significant effect of T-2 toxin (and other trichothecenes) is the immunosuppressive activity which has been clearly demonstrated in experimental animals; and which is probably linked to the inhibitory effect of this toxin on the biosynthesis of macromolecules. There is limited evidence that T-2 toxin may be carcinogenic in experimental animals.

Deoxynivalenol (DON) is probably the most widely occurring *Fusarium* mycotoxin, contaminating a variety of cereals, especially maize and wheat, in both the developed and developing world. The outbreaks of emetic (and feed refusal) syndromes amongst livestock, caused by the presence of DON in feeds, has resulted in the trivial name, vomitoxin, being attributed to this mycotoxin.

The ingestion of DON has caused outbreaks of acute human mycotoxicoses in India, China and rural Japan. The Chinese outbreak, in 1984-85, was caused by mouldy maize and wheat; symptoms occurred within five to thirty minutes and included nausea, vomiting, abdominal pain, diarrhoea, dizziness and headache.

To date, nivalenol-producing isolates of *F. graminearum* have been observed, on rice and other cereals, only in Japan and have been associated with the occurrence of red mould disease ('Akakabi-byo'). Symptoms include anorexia, nausea, vomiting, headache, abdominal pain, diarrhoea and convulsions.

Zearalenone

Zearalenone is a widely distributed oestrogenic mycotoxin occurring mainly in maize, in low concentrations, in North America, Japan and Europe. However, high concentrations can occur in developing countries, especially when maize is grown under more temperate conditions in, for example, highland regions. Zearalenone is co-produced with deoxynivalenol by *F. graminearum* and has been implicated, with DON, in outbreaks of acute human mycotoxicoses.

Exposure to zearalenone-contaminated maize has caused hyperoestrogenism in livestock, especially pigs, characterised by vulvar and mammary swelling and infertility. There is limited evidence in experimental animals for the carcinogenicity of zearalenone.

The Fumonisins

The fumonisins are a group of recently characterised mycotoxins produced by *F. moniliforme*, a mould which occurs worldwide and is frequently found in maize. Fumonisin B_1 has been reported in maize (and maize products) from a variety of agroclimatic regions including the USA, Canada, Uruguay, Brazil, South Africa, Austria, Italy and France. The toxins especially occur when maize is grown under warm, dry conditions.

The minimum water activity for the growth of *F. moniliforme* is 0.87; the maximum limit is recorded as >0.99. The minimum, optimal and maximum temperatures for growth are 2.5-5.0, 22.5-27.5 and 32-37°C, respectively. There is no information on the conditions required for the production of fumonisin B_1.

Exposure to fumonisin B_1 (FB1) in maize causes leukoencephalomalacia (LEM) in horses and pulmonary oedema in pigs. LEM has been reported in many countries including the USA, Argentina, Brazil, Egypt, South Africa and China. FB1 is also toxic to the central nervous system, liver, pancreas, kidney and lung in a number of animal species.

The presence of the fumonisins in maize has been linked with the occurrence of human oesophageal cancer in the Transkei, southern Africa and China. The relationship between exposure to *F. moniliforme*, in home-grown maize, and the incidence of oesophageal cancer has been studied in the Transkei during the ten-year period 1976-86 (Rheeder et al, 1992). The percentage of kernels infected by *F. moniliforme* was significantly higher in the high-risk cancer area during the entire period; and FB1 and FB2 occurred at significantly higher levels in mouldy maize obtained from high-risk areas in 1986.

Previously, an evaluation by the International Agency for Research on Cancer had concluded that there is sufficient evidence in experimental animals for the carcinogenicity of cultures of *F. moniliforme* that contain significant amounts of the fumonisins; whereas there is limited evidence, in experimental animals, for the carcinogenicity of fumonisin B_1.

However, the results of a recently completed study of the toxicology and carcinogenesis of fumonisin B_1 has been reported by the National Toxicology Program of the US Department of Health and Human Services. Although the report is still in draft form, it concludes that there is clear evidence of carcinogenic activity of fumonisin B_1 in male F344/N rats based on the increased incidences of renal tubule neoplasms; and that there is also clear evidence of carcinogenic activity of fumonisin B_1 in female $B6C3F_1$ mice based on the increased incidences of hepatocellular neoplasms. There is no evidence of carcinogenic activity of fumonisin B_1 in female rats or male mice.

Ochratoxin A

A. ochraceus grows more slowly than both *A. flavus* and *A. parasiticus*, but can grow at a water activity as low as 0.79. Growth has also been reported within the temperature range 8-37 °C, with an optimum variously reported as 25-31°C. Ochratoxin A is produced within the temperature range 15-37 °C, with an optimal production at 25-28 °C.

P. verrucosum grows within the temperature range 0-31°C and at a minimum water activity of 0.80. Ochratoxin A is produced over the whole temperature range. Significant quantities of toxin can be produced at a temperature as low as 4°C, and at a water activity as low as 0.86.

Exposure to ochratoxin A (OA) appears to occur mainly in wheat and barley growing areas in temperate zones of the northern hemisphere. The levels of OA reported in these commodities ranges from trace amounts to 6000 μg/kg, in

Canadian wheat. In the UK, reported levels have varied from <25 to 5,000 and from <25 to 2,700 μg/kg in barley and wheat respectively. It also occurs in maize, rice, peas, beans, cowpeas, vine fruits and their products, coffee, spices, nuts and figs.

The ability of OA to transfer from animal feeds to animal products has been demonstrated by the occurrence of this toxin in retail pork products, and the blood of swine, in Europe. Although cereal grains are considered to be the main human dietary source of OA, it has been suggested that pork products may also be a significant source of this toxin. Ochratoxin A has been found in blood (and milk) from individuals in a variety of European countries, including France, Italy, Germany, Denmark, Sweden, Poland, Yugoslavia and Bulgaria. One of the highest reported levels is 100 ng/ml OA in blood from Yugoslavia; whereas 6.6 ng/ml OA in milk has been recorded in Italy.

Existing or proposed regulations for OA are available in at least eleven countries, the permitted levels ranging from 1 to 50 μg/kg in foods and from 100 to 1000 μg/kg in feeds. In Denmark, the acceptability of pork products from a specific carcass is determined by analysing the OA content of the kidney. The pork meat and certain organs can be consumed as food if the OA content of the kidney is no more than 25 and 10 μg/kg respectively.

A provisional tolerable weekly intake of 100 ng/kg bw, of OA, approximating to 14 ng/kg body weight per day, has been recommended by a WHO/FAO Joint Expert Committee on Food Additives, JECFA.

Ochratoxin A has been linked with the human disease Balkan endemic nephropathy, a fatal, chronic renal disease occurring in limited areas of Bulgaria, the former Yugoslavia and Romania. OA causes renal toxicity, nephropathy and immunosuppression in several animal species and it is carcinogenic in experimental animals. There is sufficient evidence in experimental animals for the carcinogenicity of OA.

Patulin

Patulin is an antibiotic produced by a number of moulds. It occurs in rotten apples contaminated by *Penicillium expansum* and, consequently, may occur in apple juice and other apple-based products.

Experimental studies have demonstrated that patulin is a neurotoxin and that it produces marked pathological changes in the viscera. Although patulin has been reported as inducing local sarcomas, no mutagenic activity has been discernible in most short-term tests.

JECFA has established a provisional maximum tolerable daily intake of 400 ng/kg bw for patulin.

The Co-occurrence of Mycotoxins

The complex ecology of mould growth and mycotoxin production can produce mixtures of mycotoxins in foods and feeds, especially in cereals. The co-occurrence of mycotoxins can affect both the level of mycotoxin production and the toxicity of the contaminated material. The production of the aflatoxins in stored grains, for example, may be enhanced by the presence of trichothecenes, whereas the toxicology of naturally occurring combinations of trichothecene mycotoxins is reportedly determined by synergistic interactions, in experimental animals.

For example, in a study with swine, the affect of deoxynivalenol on weight gain and feed conversion was synergized by T-2 toxin. Interactions involving nontoxic fungal metabolites have also been reported, including the potent synergism of nontoxic *F. graminearum* metabolites (culmorin, dihydroxycalonectrin and sambucinol) with deoxynivalenol. To date, too little is known about this particularly important area of mycotoxicology.

Mycotoxins: Of Molds and Maladies

From the beginning of organized crop production through present-day agriculture, mycotoxins—toxic metabolites produced by fungi—have presented health risks to both human and animal populations. Mycotoxins are secondary metabolites produced by certain molds that infect food crops in the field and during storage.

Depending on the quantities produced and consumed, mycotoxins can cause acute or chronic toxicity in the animals and humans who eat contaminated crops or foods prepared from them. Health effects of mycotoxins may include immunological effects, organ-specific toxicity, cancer, and, in some cases, death. Agricultural workers are also at risk for dermal and respiratory exposures during crop harvest and storage. Mycotoxin contamination is a worldwide problem affecting staple crops such as corn (maize) and small grains (such as wheat), as well as tree nuts, peanuts, sorghum, and many others. Many countries regulate the maximum allowable concentrations of specific mycotoxins in food commodities and animal feed. Until recently, dietary and occupational exposures were the primary areas of concern, but with growing attention being paid to the problems associated with indoor molds and respiratory exposures, researchers are recognizing that the potential scope of mycotoxin exposures is broader than originally suspected. This scope now includes inner-city environments, with a special focus on older or poorly maintained structures that are more susceptible to harbouring molds.

Researchers are also recognizing that specific groups within a population may be more vulnerable to exposure than the population at large. For example, in the United States, the Hispanic community consumes a greater quantity of corn products compared to the general population. Because corn is vulnerable to contamination by several mycotoxins, it is possible that individuals within the Hispanic community are exposed to higher dietary levels of mycotoxins than the average American. This higher exposure could place them at a greater risk of health problems.

The effects of mycotoxins have been recorded for millennia in sources as diverse as 5,000-year-old Chinese texts, the Old Testament of the Bible, and numerous scientific journals. As research into these compounds and their effects has accumulated over the last several decades, mycotoxins have been revealed to be an extremely diverse group of compounds. During the Middles Ages, periodic epidemics of St. Anthony's fire, now known as ergotism, afflicted countless individuals who had consumed moldy rye. Gangrenous ergotism, the form that inspired the medieval name, was accompanied by fiery pain, swelling, and gangrene in the limbs. Convulsive ergotism, a second form of the toxicosis, was accompanied by convulsions and hallucinations, among other symptoms. Both forms could be fatal. Research in the 1940s identified ergot alkaloids produced by *Claviceps* spp. as the mycotoxigenic source of ergotism. Ergotism has occurred very rarely in recent decades; with regard to agricultural crops, current research focuses on the toxigenic and carcinogenic potentials associated with molds including species of *Aspergillus* and *Fusarium*. With regard to indoor molds and respiratory health issues, *Stachybotrys chartarum* (formerly *S. atra*) contributes a challenging new facet to the field of mycotoxicology.

A Centerpiece of Mycotoxin Research

"Monolithic" is how John D. Groopman, a professor of environmental health sciences at the Johns Hopkins University School of Public Health in Baltimore, Maryland, describes the literature on aflatoxins—mycotoxins produced by *Aspergillus* spp. "As far as the epidemiology in people goes, the overabundance of literature covers aflatoxins," he says. "Aflatoxin levels are regulated not only by the Food and Drug Administration [FDA] and the U.S. Department of Agriculture [USDA] but also in world commerce. I don't believe that's true for almost any of the other mycotoxins." This emphasis reflects the central role of aflatoxins, especially aflatoxin B_1 (AFB_1), in mycotoxicological research since the early 1960s.

Primarily produced by *A. flavus* and *A. parasiticus*, AFB_1 causes liver and kidney toxicity in several species and is most prominently known as a potent liver carcinogen in humans and animals. *Aspergillus* spp. exist worldwide and live off of a number of crops, although corn and peanuts are the most commonly contaminated commodities. *A. flavus* and *A. parasiticus* are the most common forms of *Aspergillus* and are also capable of producing other forms of aflatoxin. Frequently, there is contamination with multiple forms of aflatoxin. However, regardless of which aflatoxins are produced, AFB_1 is always among them, and it is the most toxic.

AFB_1 has been shown to be mutagenic in many *in vitro* model systems and is a proven carcinogen in many animal species, including rodents and nonhuman primates. The primary cancer site in these studies has been the liver, but in some cases a link has been demonstrated to lung, kidney, and colon tumours. In human epidemiological studies, chronic dietary AFB_1 exposures have been strongly linked to increased incidence of liver cancer. On the basis of this information, AFB_1 is classified as a known human carcinogen by the International Agency for Research on Cancer and the National Toxicology Program.

The carcinogenic potency of AFB_1 is not manifested until it undergoes activation by the cytochrome P450 and other oxidative enzymes. These enzymes transform the aflatoxin into several products, including the genotoxic AFB_1-*exo*-8,9-epoxide. This epoxide can be shunted by glutathione *S*-transferase in the liver, but some will intercalate, or wedge itself, between DNA base pairs. In this position, the epoxide rapidly reacts with the DNA to produce an N^7-guanyl adduct.

As demonstrated in several animal models, these adducts are produced in the greatest amounts in the liver, although some are also produced in the kidney or lung. Through DNA repair and chemical stability mechanisms, AFB_1-DNA adducts can be removed and excreted, but some adducts prove resistant to repair, thereby setting the stage for mutation events and carcinogenesis.

The potential for AFB_1-induced liver cancer is enhanced in individuals who are also infected with the hepatitis B virus, a recognized carcinogenic virus. In AFB_1-exposed populations, examination of liver tumours reveals a high incidence of a specific *p53* tumour suppressor gene mutation. In AFB_1-exposed individuals infected with the hepatitis B virus, this mutation is associated with 50-60% of the tumours. By understanding the mechanism of action of AFB_1, we are identifying guideposts for developing intervention strategies, says Groopman.

The Depth of the Field

Not all carcinogenic mycotoxins act through a genotoxic mechanism, and the fumonisins provide a case in point. The fumonisins—B_1, B_2, and B_3—are produced by *Fusarium* spp. that grow on corn, most importantly *F. moniliforme.* In horses, relatively low exposures to fumonisins have been shown to cause equine leukoencephalomalacia, a disorder characterized by brain hemorrhage and necrosis, followed by death. Horses may also suffer liver damage and possibly a degree of kidney damage following dietary exposure to fumonisins.

Liver and kidney effects are more pronounced in other species such as rodents, sheep, and rabbits. In swine, high doses of exposure to fumonisins seem to especially affect the lungs, leading to porcine pulmonary edema, a fatal condition in which fluid collects in the lungs. Low exposures result in reduced feed consumption. Early research in animals showed that fumonisins are potent cancer promoters and potentially weak initiators. The National Toxicology Program, along with the FDA's National Centre for Toxicological Research in Jefferson, Arkansas, and Centre for Food Safety and Applied Nutrition in Washington, DC, released the results of a massive study of the toxicity of fumonisin B_1 in May 1999. Their data showed that fumonisins are carcinogenic in rodents, although response differed by species and sex: male rats fed fumonisin B_1 developed liver and kidney cancers, while female mice developed liver cancer.

With regard to human toxicity, epidemiological data from southern Africa and China suggest a strong link between dietary fumonisin exposure and esophageal cancer. Human epidemiological studies, however, are not definitive, says Ken Voss, a research pharmacologist at the USDA Toxicology and Mycotoxin Research Unit in Athens, Georgia.

> *"There are suggestive data that the fungus and the fumonisins are associated with esophageal cancer," he says. "But there are enough confounding dietary and environmental factors that the correlation, although tantalizing and suggestive, is as yet far from being proven."*

According to Voss, it has been shown that fumonisins have measurable and repeatable toxic effects in animal models. Fumonisin toxicity seems to be mediated through inhibition of ceramide synthase, a key enzyme in the sphingolipid biosynthetic pathway.

"To put it in layman's terms," says Voss, "the entire metabolism of sphingolipids in the cell is disrupted." The potential ramifications of this disruption can be far-reaching, he says. Until 15-20 years ago, sphingolipids were considered as having a purely structural role in cells; however, sphingolipid molecules and their derivatives are now recognized as very biologically active compounds. These compounds, says Voss, either initiate or act as messengers for many life-or-death decisions that the cell has to make. Such decisions include whether the cell embarks on apoptosis (cell death) or enters the cell cycle and replicates. From this point, he says, there are a host of potential steps and intermediaries leading to toxigenic or carcinogenic events.

Fusarium and its Mycotoxins

In addition to fumonisins, *Fusarium* spp. produce several other mycotoxins. *F. graminearum* and *F. culmorum*, molds that contaminate corn, barley, wheat, and other crops, are capable of producing the toxins zearalenone and deoxynivalenol (also called vomitoxin). Different toxigenic species of *Fusarium* grow under different sets of climatic conditions. "The production of these compounds depends on a number of different conditions," says Retha Newbold, a supervisory research biologist at the NIEHS. "Just to have a product that is contaminated with mold is not to assume that mycotoxin is present. The mold may be there, but it may produce different levels of mycotoxins, or even different mycotoxins, depending on... different conditions."

Although zearalenone has low acute toxicity, it exhibits marked estrogenic effects in some species. Zearalenone and its metabolites, particularly-zearalenol and ß-zearalenol, have been shown to bind to estrogen receptors in experimental systems. Their estrogenic potential seems to fall between that of the endocrine-disrupting organochlorine pesticides and the more estrogenic compound diethylstilbestrol. Newbold indicates that the estrogenicity of zearalenone and its metabolites differs depending on the tissue and the species. For example, swine are especially sensitive and experience hyperestrogenism leading to reproductive problems and infertility following dietary zearalenone exposures. Other species such as cattle and sheep seem more resistant to zearalenone but may still experience some incidence of infertility, decreased milk production, and spontaneous abortion after ingesting high doses. Still other species, particularly chickens, appear even less sensitive.

It has been demonstrated that zearalenone and its metabolites may cause carcinogenesis or teratogenesis in some species, but further research is needed. Further research is also needed with regard to human toxicity. Currently, the International Agency for Research on Cancer classifies zearalenone as a 2A carcinogen, the highest possible classification when categorical human epidemiology is absent. Several countries have already established maximum allowable concentrations of zearalenone in food ranging from 0 to 1,000 micrograms per kilogram. Data on human toxicity are strongest with regard to estrogenic effects. For example, zearalenone was considered a possible etiological agent for precocious pubertal changes that were observed among Puerto Rican children for several years beginning in 1979.

Thousands of children, some of whom were shown to have zearalenone or its derivatives in their blood, reportedly experienced symptoms. However, as other estrogens (phytoestrogens or residues of animal growth promoters) were potentially present in the children's diets, this outbreak might have stronger implications with regard to zearalenone's contribution to the total environmental estrogen burden. Recent investigations using *in vitro* systems bolster the idea that zearalenone interacts with human estrogen receptors. For example, Craig Dees, a scientist in the Health Sciences Research Division of the Oak Ridge National Laboratory in Tennessee, and colleagues reported that zearalenone stimulates estrogen-receptor human breast cancer cells to enter the cell cycle *in vitro*.

> *"For people who are studying endocrine disruptors and who are actually looking for some of the potential health effects during development, this is one of their concerns," says Newbold. "But I certainly don't think it has received the attention for human health that it should have in the United States."*

With regard to research, zearalenone seems to be overshadowed by deoxynivalenol, a more demonstrably toxic *Fusarium* metabolite. This mycotoxin has been linked to large-scale poisonings, human disease, and animal production problems throughout the world. Deoxynivalenol is one of the most common mycotoxins contaminating grains. It belongs to a class of compounds called trichothecenes, to which several other mycotoxins belong. Although deoxynivalenol is the least toxic of the trichothecenes, its toxicity is still substantial in both animals and humans.

In large enough acute doses, it causes nausea, vomiting, and diarrhea and destroys blood cells. Animals, particularly pigs, demonstrate feed refusal and weight loss at lower doses.

Deoxynivalenol has also been shown to have immunological effects in animal models. For example, research reviewed by James J. Pestka, a professor of food science and human nutrition at Michigan State University in East Lansing, and colleagues in the May 1996 issue of the *Journal of Toxicology and Environmental Health* demonstrates that deoxynivalenol interferes with normal immune system functioning in mice. They concluded that deoxynivalenol induces cytokines, immune system factors that help direct an inflammatory response. In the model they reviewed, mice exposed to deoxynivalenol developed symptoms similar to human IgA nephropathy, a kidney disorder characterized by inflammation.

Expanding Frontiers in Mycotoxicology

Trichothecenes may also be at the root of an outbreak of idiopathic pulmonary hemorrhage among infants in Cleveland, Ohio. Among infants, pulmonary hemorrhage, or episodes of bleeding in the lungs, can arise from several causes such as injury or some forms of pneumonia. Unexplained, ongoing episodes of bleeding, as seen with the infant patients in Cleveland, is much rarer. Between 1993 and 1998, physicians at the Rainbow Babies & Children's Hospital in Cleveland saw 37 cases of pulmonary hemorrhage among infant patients; in the preceding 10 years, only three such cases had been encountered. Nearly all of the 37 infants were brought to the hospital because of breathing difficulties and required intensive care and ventilator support. Pulmonary hemorrhage wasn't always apparent before breathing difficulties surfaced but was detected once respiratory support began. Researchers suspected that some element in the infants' home environments was responsible for the symptoms because in several cases symptoms recurred when an infant returned to his or her home.

In a case-control investigation begun in November 1994 by the Centres for Disease Control and Prevention, researchers found evidence that the geographically clustered infants' homes were contaminated with *S. chartarum*, a mold not commonly found in home environments. *S. chartarum* is known to produce several trichothecenes, specifically satratoxins and roridin, as well as phenylspirodrimanes, cyclosporin, and a newly discovered class of compounds, the stachybocins. Researchers hypothesized that the mycotoxins, in combination with other stressors in the infants' environments such as tobacco smoke, caused the respiratory ailment that claimed the lives of 12 of the 37 infants.

> *"The primary problem with indoor molds is that the health hazard is predominantly linked to people who are atopic—that is, they tend to be allergic," says Dorr Dearborn, an associate professor of pediatrics and biochemistry at the Case Western Reserve University School of Medicine in Cleveland and one of the researchers associated with the ongoing* S. chartarum *investigation. Says Dearborn, "What we're beginning to realize more recently, which is not really well known in the medical field, is that there are molds—not just* Stachybotrys, *but probably a larger list of them—that produce mycotoxins that can have direct effects on health. This is still an area of both speculation/ conjecture and some knowledge, but it's an area of active concern and with some research at least starting to be generated."*

The precise mechanisms of the *Stachybotrys* mycotoxins are unknown, explains Dearborn. The trichothecenes may be able to trigger or aggravate an allergy problem directly, he says, but not through a traditional immunoglobulin E pathway. Typically, an allergic response involves production of antibodies constructed from IgE against the allergen and some form of inflammation (such as asthma). Pestka's work to elucidate how deoxynivalenol induces nephropathy has shown that, at low levels, the trichothecene induces inflammatory mediators. This suggests a mechanism by which mycotoxins, including those produced by *Stachybotrys*, may produce airway disease or skin reactions without going through the typical IgE mechanisms, says Dearborn.

In this literature, the effects include bleeding in the nose and throat (although not the lungs), skin irritation, and altered white blood cell counts. "The cellular mechanism of trichothecenes is well established: they are potent protein synthesis inhibitors," says Dearborn. Inhibition occurs via a single binding site on the ribosome, the cellular location of protein construction. Depending on the specific trichothecene, construction breaks down during its initiation, elongation, or termination stages, he explains. However, he continues, the details behind their effects on the immune system remain unknown.

Dearborn and his associates are currently in the early stages of a five-year grant from the NIEHS to develop an infant model for *S. chartarum* exposure as an outgrowth of their investigations into the Cleveland outbreak of idiopathic pulmonary hemorrhage. "In the Cleveland situation we're dealing with an epidemiological association—that is, we've found in the houses of the cases more than we found it in the control houses. The link is not absolute at all; it's simply an epidemiological link," says Dearborn. One of the scientists' immediate challenges comes in the form of identifying a biomarker for exposure, a difficult task given how rapidly the suspected toxins are metabolized. Finding a biomarker is difficult because suspected *Stachybotrys* toxins are rapidly metabolized and most people do not form antibodies in response to the mold. The researchers' early experiments to duplicate the disorder in infant animals have been promising. "What we have shown is that if the spores of *Stachybotrys* are instilled in the tracheas of young rat pups, they will develop pulmonary hemorrhage. Initial results suggest that we are on the proper route to develop an infant model for the disorder," says Dearborn.

Areas for Further Research

The diverse spectrum of mycotoxins produced by *S. chartarum* illustrates one of the more vexing issues in mycotoxicology: mycotoxins usually occur in mixtures. As a result, researchers recognize that interactions are possible although they are difficult to characterize. There is particular interest in exploring potential synergies, or interactions in which exposure to more than one mycotoxin results in a multiplication, rather than an addition, of risks. "The kinds of experiments that are necessary to elucidate the nature of synergy are complicated," says J. David Miller, a professor of chemistry at Carleton University in Ottawa, Canada, who recognized early on that the toxins produced by molds are

typically mixtures of toxins. Such experiments would require a lot of resources that currently are not available, so investigating potential additive effects or synergy between mycotoxins—or between mycotoxins and other environmental factors—is, for the most part, not a major research focus. Nevertheless, researchers have commented, for example, on the potential interaction between fumonisins and aflatoxin. For now, says Miller, the most important synergy that has been investigated with regard to mycotoxins is the one that exists between aflatoxin and hepatitis B.

Another area of mycotoxin research that seems ripe for further investigation concerns defining the subtler effects of individual mycotoxins. In the area of veterinary toxicology, pinning down information on such effects is an especially active area of interest, according to George Rottinghaus, a chemist in the toxicology section of the University of Missouri Veterinary Medical Diagnostic Laboratory in Columbia. "Those are what I call the gray areas," he says. "Everybody's done acute, subacute, and that type of work, but it takes a lot more effort to get into these more subtle changes. a lot of these subtle things, you really wouldn't have symptoms. It would be more of a performance-or immune-type response effect that most people wouldn't see." He offers an example: "All of a sudden your animals might be sicker than they normally are... or they off 5-10% in milk production, or they don't gain quite the way they were supposed to." Rottinghaus points out that these symptoms can be attached to many other factors, but it's hard to pin them to either mycotoxins or alternative explanations.

Controlling Exposures and Mitigating Effects

The best means of preventing the health effects of mycotoxins is to prevent exposure—a task more easily noted than achieved. In the agricultural arena, postharvest control of storage fungi is handled through proper drying and storing of grains. These measures are accomplished with varying degrees of adequacy depending on the available equipment, storage facilities, and other variables. Success in preventing field contamination can be even more variable owing to factors such as insect infestation, drought, or weather events such as hail storms. Once crops are damaged, an opening appears for fungal contamination. Whether or not fungi will exploit that opening depends on other factors, including the prevailing temperature, humidity, and water content. In some fields, microclimates may exist so that one part of the field can be heavily contaminated while neighbouring sections are completely untouched.

Although use of antifungal agents and other chemicals is potentially effective, researchers are investigating several strategies that don't rely on chemical applications. "A number of the corn companies are trying to develop hybrids that are resistant to the *Fusarium* infection," says Rottinghaus. According to Voss, researchers at the USDA are exploring another avenue in investigations centering on potential biological controls of fungal growth and toxin production in corn plants. The idea behind their strategy is to use nontoxigenic bacterial or fungal species as bioexclusion agents that would outcompete fungi in the field and in storage. This technique would only be used for corn destined for animal feed, and Voss indicates that USDA researchers anticipate some commercial applications of the technique within the next five years.

In some areas of the world, fungal control techniques are more urgently needed. Miller points out that in North America the population experiences a relatively low risk from mycotoxins owing to a diverse diet and the range of zones in which crops are raised. "In North America, we produce large amounts of a crop, and we only use a small percentage of it for human food. If we have a bad year in Texas, it's unlikely it's a bad year in Iowa. We have the luxury, by and large, to pick and choose in terms of excluding crops from our food system if it's necessary," he explains. For example, in 1996 wheat grown in Michigan, Maine, and Ontario couldn't be used because it was heavily contaminated with vomitoxin. However, because the commodity is also grown, albeit in lesser quantities, in Alberta and the U.S. Pacific Northwest, buyers could find supplies elsewhere. "In developing countries, that luxury is not there," Miller says.

In recognition of this fact, researchers are attempting to find other means of protecting populations from the health effects associated with mycotoxin exposure. Much effort has been devoted to applying such measures in AFB_1-exposed populations. One technique has been to promote vaccinations for hepatitis B in areas with high AFB_1 exposures. Another strategy explores the potential for altering the metabolism of the toxic compounds. Recently, human trials were conducted with oltipraz, a compound that interferes with the mechanism of action of aflatoxin.

The trials were published in the February 1999 issue of the *Journal of the National Cancer Institute.* "We'd like to believe from the data we have from the oltipraz clinical trial that agents that can blunt the metabolism of aflatoxin are certainly going to be important in terms of preventing aflatoxin-mediated DNA damage," says Groopman, who collaborated with researchers in China on this study. "With the oltipraz intervention, we found that we can modulate the metabolism of aflatoxins in people and shunt the metabolism toward non-DNA-damaging species. If that can be replicated by other dietary agents, that is probably going to be a very important way of intervening in large populations," he concludes.

Efforts to control exposure to mycotoxins are certainly better today than in ancient times, but they still are not perfect. For example, the testing for mycotoxins such as aflatoxin only involves grains that enter interstate commerce and thus doesn't protect people who might consume highly contaminated locally grown crops. Other populations may face greater risk simply because they consume higher-than-average amounts of certain commodities or because they live or work in poorly maintained buildings. Finally, research has primarily emphasized dietary routes of exposure. Knowledge about the long-term effects of other exposures is lacking. "The toxins that enter the crops we use for food... sort of a by-product of chemical warfare that's going on at a microbiological level," says Miller. There are lots of experiments to do, he muses, and not a lot of resources to do them.

Chilli Anthracnose Disease Caused by *Colletotrichum* Species

Colletotrichum is one of the most important plant pathogens worldwide causing the economically important disease anthracnose in a wide range of hosts including cereals, legumes, vegetables, perennial crops and tree fruits. Among these hosts, chilli (*Capsicum* spp.), an important economic crop worldwide, is severely infected by anthracnose which may cause yield losses of up to 50%. Typical anthracnose symptoms on chilli fruit include sunken necrotic tissues, with concentric rings of acervuli. Fruits showing blemishes have reduced marketability. In the *Colletotrichum* patho-system, different *Colletotrichum* species can be associated with anthracnose of the same host. Anthracnose of chilli has been shown to be caused by more than one *Colletotrichum* species including *C. acutatum* (Simmonds), *C. capsici* Butler and Bisby, *C. gloeosporioides*Penz. and Sacc., and *C. coccodes* S. Hughes.

There is little information concerning the interactions between the complexes of species involved in chilli anthracnose. This information is necessary for plant breeding purposes and disease management. Some *Colletotrichum* species respond differently to various control measures, e.g., *C. acutatum* was found to be moderately susceptible to the fungicide, benzimidazole, while *C. gloeosporioides* was highly susceptible. Correct and accurate identification will thus ultimately lead to more effective disease control and management, e.g., selecting appropriate fungicides, or long lasting resistant cultivars.

The current taxonomic status of *Colletotrichum* species is unclear. The broad host range of many species has caused problems for plant pathologists who need to identify specific plant pathogens to control disease. Molecular techniques have provided additional data to aid the naming of fungi, and have been applied to taxonomically difficult genera including *Fusarium*, *Pestalotiopsis* (Jeewon et al., 2002; 2003; 2004; Lui et al., 2007), *Mycosphaerella* and its anamorphs, but to a lesser extent to *Colletotrichum*. The management and control of anthracnose diseases are still being extensively researched. This paper reviews the host 'chilli' as well as the causal agents of the chilli anthracnose, their pathogenic variability and approaches leading to the disease management.

Host: Chilli

The genus *Capsicum* was originated in the American tropics and has been propagated throughout the world including the tropics, subtropics, and also temperate regions. The fruit of *Capsicum* has a variety of names, such as 'chilli', 'chilli pepper' or 'pepper' depending on place (i.e., differences between the English-speaking countries) and type of fruits. The term "chilli" in most of the world refers exclusively to the smaller, hot types of *Capsicum*.

Capsicum contains approximately 20~27 species, 5 of which are domesticated: *C. annuum*, *C. baccatum*, *C. chinense*, *C. frutescens*, and *C. pubescens*, and are cultivated in different parts of the world. Among the five species of cultivated *Capsicum*, *C. annuum* is one of the most common cultivated crops worldwide followed by *C. frutescens*. Chilli has many culinary advantages.

It comprises numerous chemicals including steam-volatile oils, fatty oils, capsaicinoids, carotenoids, vitamins, protein, fibre and mineral elements. Many chilli constituents are important for nutritional value, flavour, aroma, texture and colour. Chillies are low in sodium and cholesterol free, rich in vitamins A and C, and are a good source of potassium, folic acid and vitamin E. Fresh green chilli peppers contain more vitamin C than citrus fruits and fresh red chilli has more vitamin A than carrots. Two chemical groups produced by chilli are capsaicinoids and carotenoids. The capsaicinoids are alkaloids that make hot chilli pungent. A large number of carotenoids provide high nutritional value and the colour to chilli.

Constraints to Chilli Production

Chilli is considered to be one of the most important crops in the tropics. The area cultivated with chilli worldwide is about 1 700 000 ha for producing fresh chilli, and around 1 800 000 ha for producing dried chilli; a total area of 3 729 900 ha with a total production of 20 000 000 t. The most important producers and exporters of chilli include China, India, Mexico, Morocco, Pakistan, Thailand and Turkey.

Diseases caused by fungi, bacteria and viruses are the major constraints to chilli production. Anthracnose disease caused by *Colletotrichum* species, bacterial wilt caused by *Pseudomonas solanacearum*, and mosaic disease caused by chilli veinal mottle virus (CVMV) or cucumber mosaic virus (CMV) are the most serious destructive diseases of chilli. Anthracnose disease caused by *Colletotrichum* species is one of the most economically important diseases reducing marketable yield from 10% to 80% of the crop production in some developing countries, particularly in Thailand.

Anthracnose is mainly a problem on mature fruits, causing severe losses due to both pre-and post-harvest fruit decay.

Anthracnose Disease

Anthracnose, derived from a Greek word meaning 'coal', is the common name for plant diseases characterized by very dark, sunken lesions, containing spores. Generally, anthracnose disease is caused by *Colletotrichum* species which belongs to the Kingdom Fungi; Phylum Ascomycota, Class Sordariomycetes; Order Phyllachorales; and Family Phyllachoraceae.

The anamorphs are *Glomerella* species. Anthracnose of chilli was first reported from New Jersey, USA, by Halsted (1890) in 1890 who described the causal agents as *Gloeopsorium piperatum* and *Colletotrichum nigrum*. These taxa were then considered as synonyms of *C. gloeosporioides* by von Arx (1957).

Anthracnose causes extensive pre-and post-harvest damage to chilli fruits causing anthracnose lesions. Even small anthracnose lesions on chilli fruits reduce their marketable value. Many post-harvest diseases of fruit exhibit the phenomenon of quiescence in which symptoms do not develop until the fruit ripens.

Colletotrichum species are the most important pathogens that cause latent infection. Appressoria are known to form adhesive disks that adhere to plant surfaces and remain latent until physiological changes occur in fruits. Appressoria that formed on immature fruits may remain quiescent until ontogenic changes occur in the fruits. Anthracnose disease can occur on leaves, stems, and both pre-and post-harvest fruits.

Typical fruit symptoms are circular or angular sunken lesions, with concentric rings of acervuli that are often wet and produce pink to orange conidial masses. Under severe disease pressure, lesions may coalesce. Conidial masses may also occur scatteredly or in concentric rings on the lesions.

Many studies have concluded that disease management practices are often inadequate to eliminate the diseases. Breeding to develop the long-lasting resistant varieties has also not been successful due to involvement of multiple *Colletotrichum* species in anthracnose infection.

Causal Agents of Chilli Anthracnose

In the *Colletotrichum* patho-system, different *Colletotrichum* species can be associated with anthracnose of the same host. *Colletotrichum* species causing anthracnose of chilli have been reported from different countries and regions. Although these species have been the subject of numerous investigations, there remain many gaps in the knowledge of the disease process and understanding of the complex relationships between the species involved. Kim et al.(2004) reported that different species cause diseases of different organs of the chilli plant; for example, *C. acutatum* and *C. gloeosporioides* infect chilli fruits at all developmental stages, but usually not the leaves or stems, which are mostly damaged by *C. coccodes* and *C. dematium*. Leaf anthracnose of chilli seedlings caused by *C. coccodes* was first reported in chilli growing in a field in Chungnam Province of Korea in 1988. Different *Colletotrichum* species may also play an important role in different diseases of mature stages of chilli fruit as well.

For example, *C. capsici* is widespread in red chilli fruits, whereas *C. acutatum* and *C. gloeosporioides* have been reported to be more prevalent on both young and mature green fruits. Anthracnose caused by *C. coccodes* does not result in severe epidemics on chilli fruits. *C. gloeosporioides*, the predominant species on chilli in Korea, was differentiated into G and R strains by isozyme analysis of esterase, leucine amino peptidase, phosphatase and glutamine oxalocetic trasminase.

Colletotrichum species can survive in and on seeds as acervuli and micro-sclerotia. Survival of mycelia and stomata in colonized chilli seeds had been reported. It has been shown that the pathogen readily colonizes the seed coat and peripheral layers of the endosperm even in moderately colonized seeds. Heavily colonized seeds had abundant inter-and intracellular mycelia and acervuli in the seed coat endosperm and embryo, showing disintegration of parenchymatous layers of the seed coat and depletion of food material in endosperm and embryo.

Fungi can overwinter on alternative hosts such as other solanaceous or legume crops, plant debris and rotten fruits in the field. *Colletotrichum* species naturally produce micro-sclerotia to allow dormancy in the soil during the winter or when subjected to stressful conditions, and these micro-sclerotia can survive for many years. During warm and wet periods, conidia from acervuli and micro-sclerotia are splashed by rain or irrigation water from diseased to healthy fruit and foliage. Diseased fruit acts as a source of inoculum, allowing the disease to spread from plant to plant within the field.

Initial infection by *Colletotrichum* species involves a series of processes including the attachment of conidia to plant surfaces, germination of conidia, production of adhesive appressoria, penetration of plant epidermis, growth and colonization of plant tissue and production of acervuli and sporulation. Anthracnose is mainly a problem on mature fruits, causing both pre-and post-harvest fruit decay resulting severe economic losses. Appressoria that formed on immature fruits may remain quiescent until the fruits mature or ripen.

Disease Cycle and Epidemiology of Anthracnose

Environmental factors play a major role in the development of disease epidemics. The relationships among rainfall intensity, duration and crop geometry and the dispersal of inoculum possibly lead to different levels of disease severity. The effects of temperature often interact with other factors, such as leaf surface wetness, humidity, light or competitive microbiota.

The duration of the surface wetness, however, appears to have the most direct influence on the germination, infection and growth of the pathogen on the host. Generally infection occurs during warm, wet weather. Temperatures around 27 °C and high humidity (a mean of 80%) are optimum for anthracnose disease development.

Colletotrichum species utilize diverse strategies for invading host tissues, which vary from intracellular hemibiotrophy to subcuticular intramural necrotrophy. *Colletotrichum* species produce a series of specialized infection structures such as germ tubes, appressoria, intracellular hyphae, and secondary necrotrophic hyphae. These pathogens infect plants by either colonizing subcuticular tissues intramurally or being established intracellularly. The preinfection stages of the both are very similar, in which conidia adhere to and germinate on the plant surface, producing germ tubes that form appressoria which in turn penetrate the cuticle directly. Following penetration, the pathogens that colonize the intramural region beneath the cuticle invade in a necrotrophic manner and spread rapidly throughout the tissues.

There is no detectable biotrophic stage in this form of parasitism. In contrast, most anthracnose pathogens exhibit a biotropic infection strategy initially by colonizing the plasmalemma and cell wall intracellularly. After the biotrophic state, intracellular hyphae colonize one or two cells and subsequently produce secondary necrotrophic hyphae.

These pathogens are therefore regarded as hemibiotrophs or facultative biotrophs. For example, *C. gloeosporioides* on avocado, chilli and citrus can produce both types of colonizations: intracellular biotrophy at an early stage and intramural necrotrophy later.

Although the mechanisms developed by *Colletotrichum* species appear similar in prepenetration events, there are differences between species in the later mechanisms such as spore adhesion, melanization and cutinization in penetration of the plant cuticle by the appressoria. For example, the host-pathogen interaction of *C. acutatum* appears to be more biotrophic than that of some other species such as *C. gloeosporioides*. Based on studies with *C. acutatum* on specific hosts, four types of interactions or infection strategies were described by Peres et al.(2005) as follows:

1. Biotrophic growth of *C. acutatum* with secondary conidiation in which conidia germinate to form appressoria and quiescent infections, and secondary conidia are formed after germination of the appressoria (e.g., predominantly biotrophic disease cycle on citrus leaves).
2. Subcuticular intramural necrotrophy with hyphal development within periclinal and anticlinal walls of epidermal host cells which are swollen and wider apart (e.g., predominantly necrotrophic disease cycle on strawberry).
3. Hemibiotrophic interaction with infection vesicles and broad primary hyphae within host cells. Inter-and intracellular hyphal growth could be seen as the subsequent necrotrophic phase (e.g., combination of

biotrophy and necrotrophy but mostly a biotrophic disease cycle on blueberry fruits).

4. Hemibiotrophic and subcuticular, intra-and intercellular development of *C. acutatum* (e.g., combination of biotrophy and necrotrophy but mostly a necrotrophic disease cycle on almond leaves and fruits).

There are only a few detailed studies on penetration and colonization by *Colletotrichum* species on chilli. Kim et al.(2004) noticed that there was no biotrophic infection vesicle found during the infection process of *C. gloeosporioides* in susceptible chilli. Epidermal cytoplasm became condensed and small vacuoles increased and cell destruction extended to the subepidermal cells of the plant, which are likely to be damaged by the pathogen enzymes. At later stages of infection, tissues were colonized inter-and intracellularly by the pathogen. This structural feature indicated that the infection was governed by necrotrophic fungal growth.

Colletotrichum species are generally able to survive in or on seeds and one of the ways that anthracnose is introduced to the chilli field is through infected transplants. *C. capsici* infection of chilli was shown to have two pathways: invasion through the seed coat and invasion through the openings of the testa. Jewsakun (1978) mentioned that *C. capsici* caused root rot of seedlings; however, whether chilli anthracnose can be seed-transmitted, and the role of seed infection and seedling infection in pre-and post-emergence damage of chilli plants are still questionable. We have found that *C. acutatum* can infect chilli seeds either by reducing the germination rates or by causing damping off of seedlings. However, very little is known about the disease cycle of the pathogens that cause anthracnose in chilli. There are still several questions to be answered.

Identification of *Colletotrichum* Species

Conventional Methods

Accurate identification of *Colletotrichum* species along with the knowledge of populations responsible for epidemics are essential for developing and implementing effective disease control strategies. Traditionally, identification and characterization of *Colletotrichum* species have been based on morphological characters, such as size and shape of conidia and appressoria; existence of setae; the teleomorph state and cultural characters such as colony colour, growth rate and texture. These criteria alone, however, are not always adequate for species identification due to overlap in morphological characters and phenotypic variation among species under different environmental conditions.

Conidial shape has been applied as a reliable means of discriminating certain species; for example, conidial shape has differentiated *Colletotrichum* species pathogenic to strawberry. However, in other cases, identification can be complicated because of overlapping ranges of conidial morphology and variation in colony characteristics. Correct taxonomic identification is important in disease management such as choosing appropriate fungicides. For instance, it could be seen in different responses of *C. acutatum* and *C. gloeosporioides* to benzimidazole-based fungicides.

Molecular Genetics Approach

To overcome the inadequacies of traditional morphology based identification schemes, DNA sequence analyses have been used to characterize and analyse the taxonomic complexity of *Colletotrichum*. Cannon et al.(2000) stated that data derived from nucleic acid analyses should provide the most reliable framework to build a classification of *Colletotrichum*, as DNA characters were not directly influenced by environmental factors.

Most fungal phylogenetic studies utilized sequences from the ribosomal gene cluster, since they were present in large numbers as tandem repeats and evolved as a single unit. In particular, sequence analysis of the internal transcribed spacer (ITS) regions which lie between the 18S and 5.8S genes and the 5.8S and 28S genes, has proved useful in studying phylogenetic relationships of *Colletotrichum* species because of their comparative variability. Apart from ITS region, sequence analysis of protein coding genes such as partial â-tubulin gene, has also been applied to resolve phylogenetic relationships among *C. acutatum* species complexes. Sequences of introns from two genes (glutamine synthase and glyceraldehyde-3-phosphate dehydrogenase) were also used to evaluate a diverse collection of isolates of *C. acutatum*. *C. acutatum* isolates clustered into groups. These groups might represent phylogenetically distinct species of *C. acutatum sensu lato*. Yun et al.(1999) stated that, because of the high intra-species variability and the low inter-species variability, MAT1-2 mating type sequences gave strong support for branches, allowing differentiation of closely related *Cochliobolus* spp. whose relationships were not resolved by ITS sequences alone. Consequently, Du et al.(2005) confirmed that MAT1-2 mating type was useful in differentiating the groups of isolates from the species complexes (*C. graminicola*, *C. gloeosporioides* and *C. acutatum*). However, there is no report concerning the use of these genes to differentiate between the *Colletotrichum* species involved in chilli anthracnose.

A combined application of molecular diagnostic tools along with traditional morphological characterization is an appropriate and reliable approach for studying *Colletotrichum* species complexes. Than et al.(2008) differentiated isolates of chilli anthracnose from Thailand into three species: *C. acutatum*, *C. capsici* and *C. gloeosporioides*, based on morphological characterization, sequencing based on rDNA-ITS region and partial beta tubulin gene and pathogenicity testing. Hong and Kim (2007) reported that Korea isolates of *C. acutatum* were phylogenetically separated from the global groups of *C. acutatum* A1 to A8 based on the sequences in partial beta-tubulin 2 (exons 3-6). Restriction fragment length polymorphisms (RFLP) of ITS region resulting from *Alu*I, *Rsa*I and *Bam*HI digestions have also been employed to differentiate *Colletotrichum* species from chilli anthracnose in Taiwan region. Four species of *Colletotrichum* were identified by ITS-RFLP fingerprinting and observation of undistinguishable isolates of *Colletotrichum* from their studies indicated the various inter-and intra-species variations in *Colletotrichum* species. Ratanacherdchai et al.(2007) distinguished *C. capsici* and *C. gloeosporioides* causing chilli anthracnose in Thailand using random amplified polymorphic DNA (RAPD) markers.

Pathogenic Variability of *Colletotrichum* Species

When any of the progeny exhibits a characteristic that is different from those present in the ancestral individuals or descent individuals, this individual is called a variant. This may involve a change in any conceivable biological characteristic, such as colour, shape, growth rate and reproduction rate. In the case of pathogens, changes in host range may occur, i.e., it may be able to infect a variety of the host plant or cultivar not previously infected by the ancestor, or in virulence, i.e., it may produce a milder or much more severe disease than the ancestors. This is the way that resistance of a plant variety is 'broken down'. The spread of a resistant genotype capable of escaping a current prevalent pathogen will be challenged by a new parasitic strain that harbours a virulent gene which is capable of overcoming that resistance. Compatibility of plant-pathogen interactions is often governed by the gene-for-gene model in many pathosystems. This suggests a continuous co-evolutionary change in both host and parasite. Some pathogen populations are known to be pathogenitically diverse, and the diversity seems to be due to continuous generation of novel pathogenic variations. Information on pathogen diversity and the geographic distribution of the pathogen population is therefore a prerequisite for accurate assessment of durable resistant germplasm in breeding programs. Taylor and Ford (2007) suggested that knowledge of pathotype diversity is important when choosing the appropriate isolates to screen for resistance in plant breeding programs.

We have studied pathogenic variation among 10 isolates of *C. acutatum* against 7 cultivars including reportedly susceptible species of *Capsicum* (*Capsicum annuum* 'Bangchang') and resistant species such as *Capsicum chinense* 'PBC 932', *Capsicum baccatum* 'PBC 80' and 'PBC 81' in Thailand. This revealed two pathotypes based on qualitative differences in infection of a reported resistant genotype, *C. baccatum* genotype 'PBC 81'. Interestingly, of 10 isolates assayed, the genotype showed complete resistance against 5 isolates tested, whereas it showed a highly susceptible reaction to the other 5 isolates tested. Sharma et al.(2005) studied the pathogenic variability in *C. capsici* in India and proposed that 15 pathotypes of *C. capsici* existed among 37 isolates from different chilli growing regions from Himachal Pradesh in India.

However, these pathotype differences were based on quantitative differences in host reaction, i.e., level of aggressiveness. Taylor et al.(2007) defined the pathotype as "a subclass or group of isolates distinguishable from others of the same species by its virulence on a specific host (genotype), i.e., a qualitative difference in disease severity" and the aggressiveness as "the natural variation in virulence or level of disease (measured quantitatively) within the pathogen population." According to Taylor and Ford (2007), there might be confusion as to whether true pathotype differences exist, or whether the differences observed in disease severity are a measure of the natural distribution of aggressiveness within a population, ranging from low to high. However, the level of aggressiveness of isolates is also an important consideration in resistance breeding programmes and disease control management.

Several studies have screened *C. acutatum*, which is the very virulent species, against chilli genotypes and found that *Capsicum baccatum* genotype 'PBC 80' is a genetic resource pool for resistance to anthracnose. We have also confirmed this. This genotype is assumed to be useful for studying genetics of resistance and practical breeding in chilli pepper. However, we found that another genotype of *C. baccatum*, 'PBC 81', showed high susceptibility to some *C. acutatum* isolates. This would make it questionable as to whether *C. baccatum* species are a useful genetic resource pool for breeding for resistance to anthracnose. In contrast to *C. baccatum*, susceptibility of the *C. annuum* cultivars has been reported in several studies. In addition, *Capsicum chinense* 'PBC 932' has been reported as a resistant variety to *Colletotrichum capsici* and hence has been introgressed with *C. annuum* 'Bangchang' to produce F_1 progeny. However, Yoon et al.(2004) and Than et al.(2008) found that the 'PBC 932' was highly susceptible to *C. acutatum* isolates. Nevertheless, this information about highly susceptible reaction of 'PBC 932' will help chilli breeders to be aware of the potential for breaking down its resistance to *C. acutatum*.

Population Structure and Molecular Marker Techniques

The genetic structure of plant pathogen populations, the potential for gene flow and long distance dispersal, and the relative contribution of sexual and asexual reproduction have direct inference with agricultural ecosystems. Abang (2003) stated that genetic population structure refers to the amount and distribution of genetic variation within and between populations. For example, how rapidly a pathogen can evolve could be indicated by the amount of genetic variation maintained within a population and then which may eventually be used to predict how long a control measure is probable to be effective. Pathogens with large genetic variations are assumed to adapt faster to fluctuating environments.

In general, organisms reproduce by means of a sexual process. Variation in progeny is introduced primarily through segregation and recombination of genes in meiotic division of the zygote. However, in the case of anamorphic fungi such as *Colletotrichum*, reproduction is mainly or exclusively vegetative. Parasexual reproduction, by which a system of genetic recombination can occur within fungal heterokaryon, will likely lead to variation. The new variants come into existence, which may be identical in appearance to that of the ancestral types, but behave differently as far as disease production is concerned.

In recent decades, molecular markers have been widely used to measure the variation in a pathogen population. Random amplified polymorphic DNA (RAPD), simple sequence repeat (microsatellite), inter simple sequence repeat (ISSR), amplified fragment length polymorphism (AFLP), and genomic DNA RFLP, are among the most frequently used neutral genetic markers in population genetic studies. Despite its importance, there have only been a few studies concerning the diversity of *Colletotrichum* species associated with chilli anthracnose. Sharma et al.(2005) found a variable population of *C. capsici* causing fruit rot/die back or anthracnose of chilli in the northwestern region of India based on differential inoculation tests and RAPD analysis.

In contrast, examination of *C. acutatum* from a limited collection of isolates from chilli anthracnose epidemics in Brazil, Korea, Taiwan region and the US indicated that they belonged to a single group based on mtDNA-RFLP analysis. They also found that most isolates (36 out of 43) examined from the various countries and regions belonged to a single vegetative compatibility group (VCG). As the sexual stages of *C. acutatum* have never been found in nature, the asexual reproduction mode of *C. acutatum* is likely to be an important role in its population structure.

Chapter 5

The Growth Hormone in Plants

Antiauxins are another class of synthetic auxin analogs. These compounds such as x-(p-chlorophenoxy)isobutyric acid, or PCIB, have little or no auxin activity but specifically inhibit the effects of auxin. When applied to plants, antiauxins may compete with IAA for specific receptors without triggering an auxin response, thus inhibiting normal auxin action. It is possible to overcome the inhibition of an antiauxin by adding excess IAA.

How can such a diverse group of chemicals all be active auxins? No other plant hormone, with the possible exception of the cytokinins, has such a wide variety of active analogs. For example, few people would have predicted that either 2,4-dichlorophenoxyacetic acid (2,4-D) or α-naphthalene acetic acid (α-NAA), both of which bear little resemblance to IAA, would be able to stimulate coleoptile elongation or have activity in other auxin bioassays.

A comparison of the compounds that possess auxin activity reveals that at neutral pH they all have a strong negative charge on the carBoxyl group of the side chain that is separated from a weaker positive charge on the ring structure by a distance of about 0.5 nm. This charge separation may be an essential structural requirement for auxin activity. The main part of the molecule, the indole ring, is not essential for activity, although an aromatic or fused aromatic ring of a certain size range is required. Recently Edgerton and colleagues (1994) proposed a set of molecular requirements for auxin activity based on studies of the binding of various auxin analogs to a protein, auxin-binding protein 1 (ABP1), which may be the auxin receptor.

Their model defines three essential regions of the binding site: a planar aromatic ring–binding platform, a carBoxylic acid–binding site, and a hydrophobic transition region that separates the two binding sites.

Auxin Measurement by Radioimmunoassay

A radioimmunoassay (RIA) allows the measurement of physiological levels (10^{-9} g = 1 ng) of IAA in plant tissues. An RIA, requires two chemicals: (1) specific antibodies that recognize IAA; and (2) a radioactively labelled IAA in which several hydrogen atoms have been replaced by radioactive tritium. Because of the sensitivity and selectivity of the antibodies used in this technique, RIA is as sensitive as some of the best physicochemical methods. Radioimmunoassays also allow one to quantify a specific auxin. In a modified RIA, the radiolabeled IAA is replaced by an auxin conjugated to an enzyme, generally an alkaline phosphatase.

This enzyme-linked immunosorbent assay (ELISA) affords sensitivities that are similar or better than the RIA.

Evidence for the Tryptophan-Independent Biosynthesis of IAA

Further evidence was provided by labelling studies with tryptophan auxotrophic mutants of *Arabidopsis*. Together, these studies established that the branch point for IAA biosynthesis is either indole or its precursor, indole-3-glycerol phosphate. Studies with the *Arabidopsis* mutants *trp2* and *trp3*, which are blocked in the last two steps of tryptophan biosynthesis, have shown that IAN accumulates up to 11-fold in the mutants compared to the wild type.

In tomato, IPA has been shown to be synthesized independently of tryptophan. Depending on the species, then, either IAN or IPA may serve as the intermediate between either indole-3-glycerol phosphate or indole, respectively, and IAA. The discovery of the tryptophan-independent pathway has drastically altered our view of IAA biosynthesis, but the relative importance of the two pathways (tryptophan-dependent versus tryptophan-independent) is poorly understood. Some plants, such as bean seedlings, synthesize IAA primarily by the tryptophan-dependent pathway. The same is true of embryogenic carrot cell suspension cultures. However, upon removal from the medium containing 2,4-D, which initiates embryogenesis in these cultures, the cells switch over to the tryptophan-independent pathway of IAA biosynthesis. This result suggests that the specific pathway of IAA biosynthesis that is utilized may be under developmental control.

The Multiple Factors That Regulate Steady-State IAA Levels

The steady-state level of free IAA in the cytosol is tightly regulated and is determined by several interconnected processes. Auxin metabolism includes the synthesis, degradation, and conjugation of IAA.

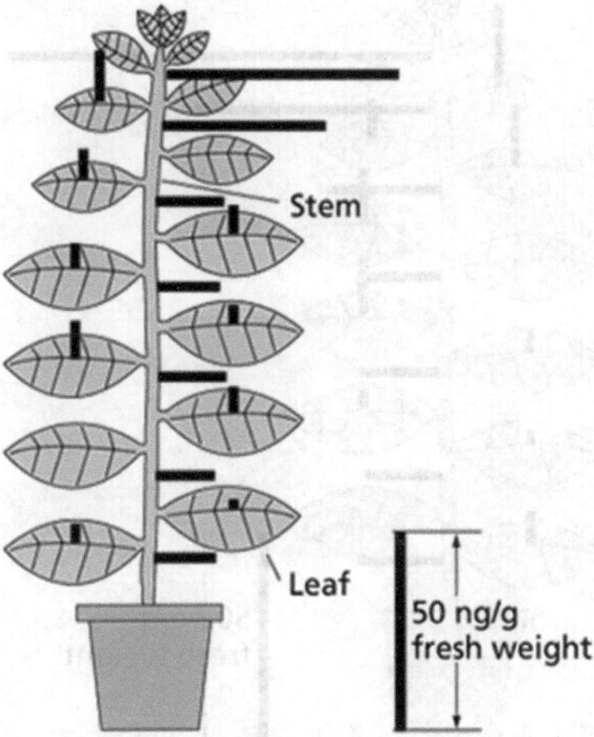

Figure: *Concentrations of IAA in different regions of the shoot of a wild-type tobacco plant. The bars at the different regions indicate the auxin concentrations found in leaves and stems.*

In addition to IAA metabolism, its compartmentation in the choroplast and transport also play important roles in regulating the level of free IAA. The amounts of IAA in different regions of a normal tobacco plant. The highest concentration of free IAA is in the apical bud. The level of free IAA in the mature regions of the stem is nearly constant. Since there is little IAA synthesis in the mature regions of the shoot, we can infer that there must be little catabolism or conjugation in these regions; otherwise the free IAA concentrations would continue to decline.

What happens when IAA is produced uniformly at high levels throughout the plant? To find out, tobacco plants were transformed with the two IAA biosynthesis genes from *Agrobacterium tumefaciens* so that IAA was produced throughout the plant.

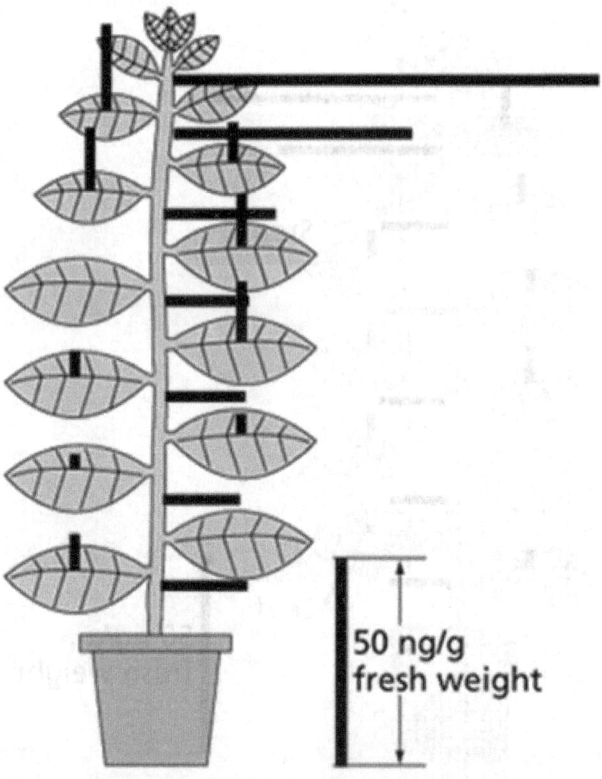

Figure: *Concentrations of IAA in different regions of the shoot of a transformed tobacco plant expressing the bacterial genes that encode the two enzymes involved in the IAM pathway of IAA biosynthesis: Trp monoxygenase and IAM hydrolase. The bars at the different regions indicate the auxin concentrations found in leaves and stems.*

The level of free IAA in the upper part of the shoot increased considerably, while the free IAA concentrations in the mature parts of the shoot hardly changed. The measured levels of conjugated IAA were several times higher in the transformed tobacco plant than in the control. We can infer that processes that decrease the amount of free IAA must have been activated in the mature shoot so that the concentration of free IAA was maintained close to its normal level.

The Mechanism of Fusicoccin Activation of the Plasma Membrane H^+-ATPase

Our examination of the auxin response would be incomplete without a discussion of the mechanism of fusicoccin (FC). The structure of fusicoccin A, a phytotoxin produced by the fungus *Fusicoccum amygdali*. This phytotoxin is produced by the fungus *Fusicoccum amygdali*, a parasite of peach and almond trees. Once released into the leaves, the toxin stimulates the H^+-ATPases of the guard cells. At the biochemical level it induces a rapid (10 to 30 seconds) hyperpolarization of the plasma membrane, accompanied by an acidification of the cell wall. This triggers an irreversible opening of the stomata resulting in wilting of the leaves and, eventually, the death of the tree.

FC causes membrane hyperpolarization and proton extrusion in nearly all plant tissues. Treatment of coleoptiles and stem sections with FC leads to a transient growth response, a result that provides support for the acid growth hypothesis for auxin-induced growth. But whereas auxin-induced proton extrusion has a lag time of about 10 minutes and is inhibited by cycloheximide, FC-induced proton extrusion begins after only 1 to 2 minutes, and the response is insensitive to cycloheximide. Moreover, FC-treated cells can acidify down to a much lower extracellular pH than auxin-treated cells can.

Because of these effects, FC has sometimes been referred to as a "super auxin." However, FC acts by a mechanism entirely different from that of auxin. For example, FC does not stimulate the expression of any of the auxin-induced genes, nor can it mimic the effects of IAA on other developmental processes, such as cell division. Thus the effects of FC on plants are much more limited than those of IAA, which is not surprising since FC is a toxin, not an endogenous plant hormone. In the case of proton extrusion, FC is known to activate preexisting H^+-ATPases on the plasma membrane, probably by displacing the autoinhibitory C-terminal domain of the enzyme from the catalytic site. In contrast, auxin may promote proton extrusion in part by activation of preexisting H^+-ATPases (perhaps via acidification of the cytosol) and in part by causing the de novo synthesis of the H^+-ATPase.

Fusicoccin was first identified as a phytotoxin in the 1960s, but it was not until 1977 that a radioactively labelled version of this toxin was available that could be used to characterize its ligand. Studies with radioactively labelled FC have shown that it binds tightly to a protein on the plasma membrane. Ultimately, these studies led to the purification and identification of the fusicoccin-binding protein (FCBP).

Initial attempts to purify FCBP after photoaffinity labelling led to conflicting results. Some laboratories identified a homodimer, composed of two 30 kDa subunits. The ability of the purified 30 kDa polypeptides to stimulate the plasma membrane H^+-ATPase in the presence of FC was tested in vitro proton-pumping experiments. The 30 kDa polypeptide was reconstituted along with purified plasma membrane ATPase into artificial membrane vesicles called proteoliposomes. The addition of FC to the reconstituted proteoliposomes caused an increase in their ATP-driven H^+-pumping activity. Cloning and sequencing of the gene for the 30 kDa polypeptide revealed it to be a member of the so-called 14-3-3 family of regulatory proteins. Regulatory proteins of the 14-3-3 family are widespread among plants and animals. Originally identified in mammalian brain tissue (and named after the positions they occupy in chromatographs), 14-3-3 proteins make up a heterogeneous family of soluble proteins ranging from 25 to 32 kDa, which associate to form dimers. The precise function of 14-3-3 proteins has not yet been elucidated. However, the numerous functions that have been attributed to them indicate that they play multiple roles in signal transduction pathways.

Although it was initially believed that the 30 kDa 14-3-3 protein represented the complete FCBP, another laboratory reported that a 90 kDa polypeptide consistently copurified with the two 30 kDa polypeptides. More recently it has been demonstrated that the 14-3-3 protein is unable to bind FC by itself. To bind FC, the 14-3-3 protein requires the presence of H^+-ATPase. This finding confirms the earlier report that a 90 kDa polypeptide (the H^+-ATPase) copurifies with the 14-3-3 protein. According to the current model, FC activates the H^+-ATPase by stabilizing a transient complex that forms between the autoinhibitory C-terminal domain of the H^+-ATPase and the 14-3-3 dimer locking the enzyme into its most active state.

The Fluence Response of Phototropism

Phototropism involves two fundamental processes: blue-light perception and differential cell elongation. The identity of the blue-light photoreceptor involved in phototropism. There is now compelling evidence that two flavoproteins associated with the plasma membrane (PHOT1 and PHOT2) are the photoreceptors for phototropism. These proteins become phosphorylated in response to blue light. The differential cell elongation in response to unilateral light is thought to be caused by the lateral redistribution of auxin to the shaded side.

If phototropism were a simple photochemical reaction, one would expect the fluence response curve to rise from a threshold level of light to a plateau at which the photoreceptor becomes saturated. However, this is not what happens. The figure shows a diagram of a typical fluence response curve, which is similar to what has been observed in a wide range of species, including *Avena* coleoptiles and *Arabidopsis* stems.

Starting at the light threshold, there is a bell-shaped curve that has been termed the first positive curvature. The first positive curvature is followed by a neutral zone in which little or no measurable response is observed. Under certain conditions, *Avena* coleoptiles may actually bend away from the light in this range, a phenomenon referred to as the first negative curvature. Both the first positive and the first negative curvatures (neutral zone) are restricted to the coleoptile tip. At higher light fluences the curve ascends again, and this second bending response is termed the second positive curvature. The second positive curvature differs from the first positive curvature in that the bending occurs at the base rather than at the tip of the coleoptile.

Plant physiologists have long puzzled over the meaning of this complex relationship between light fluence and phototropism. Most studies of phototropism have been carried out in the range of the first positive curvature. The first positive curvature and the neutral zone obey the Bunsen–Roscoe reciprocity law, which states that the response will be the same for a brief exposure to bright light as for a longer exposure to dim light, as long as the product of time and the photon fluence rate are the same. Thus the first positive curvature is a response to short light pulses.

In contrast to the first positive curvature, the second positive curvature is a response to irradiations of long duration. The second positive curvature does not obey the reciprocity law; instead, it is proportional to the duration and the photon fluence rate. For the second positive response to occur, both the time threshold and the fluence threshold must be exceeded. Since these are the conditions that prevail normally in nature, phototropism in nature usually represents the second positive curvature.

What causes the gradient in auxin to form at the three fluences? It has long been assumed that the light gradient that is established across the coleoptile during unilateral irradiation with blue light must generate a biochemical gradient between the cells of the shaded and the irradiated sides, but the nature of this biochemical gradient has, until recently, eluded detection. With the discovery of blue light–induced phosphorylation of the PHOT1 and PHOT2 plasma membrane proteins, much work has focused on these photoreceptors.

Recently, Michael Salomon and his colleagues at the University of Munich demonstrated that unilateral blue light causes a lateral gradient in the phosphorylation of PHOT1 from the irradiated to the shaded side. Moreover, the concentration of PHOT1 declines exponentially from the tip of the coleoptile to the base. Studies on the rates of phosphorylation of the 116 kDa protein PHOT1 at the tip versus at the base of oat coleoptiles have led to a new model for the fluence response curve of phototropism.

According to this new model proposed by Salomon and colleagues, the gradient in concentration of NPH1 from tip to base makes the tip more sensitive to light than the base is. Thus, at threshold fluences, a gradient in phosphorylation occurs only at the tip, leading to a small bending response. Increasing the fluence increases the gradient of phosphorylation at the tip, until phosphorylation on the irradiated side saturates and a maximum phosphorylation gradient is obtained. This maximum gradient represents the peak of the first positive curvature.

As the fluence is increased beyond this point, phosphorylation increases on the shaded side due to the penetration of light, reducing the gradient. When phosphorylation of NPH1 is equal on both sides of the tip, no bending occurs (the neutral zone).

However, with continued exposure to light of sufficient fluence, phosphorylation increases at the base on the irradiated side. Phosphorylation at the base is dependent on both fluence and time, unlike phosphorylation at the tip which is strictly dependent on fluence. Thus, phosphorylation at the base appears to involve some type of time-dependent biochemical process. Since recent immunological studies have shown that the total amount of the 116 kD protein remains constant during phototropism, the time-dependent increase in phosphorylation at the base does not involve synthesis of new PHOT1 protein. Instead, a second photoreceptor (PHOT2), initially implicated in chloroplast movement in response to light, appears to function as an additional photoreceptor in second positive curvature conditions. The model proposed by Salomon and colleagues requires further testing, and other interpretations are certainly possible. Nevertheless, the new model provides the first plausible explanation for the fluence response curve of phototropism. However, even if this model is confirmed, the question remains as to how a gradient in phosphorylation can bring about a gradient in auxin.

Gravity Perception without Statoliths in *Chara*

An alternative mechanism of gravity perception that does not involve statoliths has been documented in the giant-celled fresh water alga *Chara*. Despite the absence of any discernible sedimenting statoliths, *Chara* filaments show negative gravitropism. To explain this phenomenon, it has been suggested that the entire protoplast acts like a statolith, and that gravity is perceived at the interface between plasma membrane and cell wall.

According to the Plasmalemma Central Control (PCC) model, stretch-activated calcium channels are clustered around attachment centres connecting the cytoskeleton and the cell wall. These calcium channels are thought to open in response to tensions at the membrane caused by gravity-induced changes in the distribution of forces exerted by the protoplast, the cytoskeleton, or the cell wall. Something like the PCC model may explain how *Chara* and the starchless mutants of *Arabidopsis* are able to perceive and respond to gravity.

Cytokinins: Regulators of Cell Division

Cultured Cells Can Acquire the Ability to Synthesize Cytokinins

When cultured normal callus tissues of many species are subcultured repeatedly over a long period, they may become hormone autonomous; that is, they will grow on culture medium lacking an auxin or a cytokinin. This phenomenon is known as habituation. There are several possible explanations for the habituation of plant tissues. First, the tissues may acquire the capacity to synthesize their own cytokinin, and in some cases habituated callus tissues have been shown to contain cytokinins. Alternative explanations are that the rate of cytokinin destruction is reduced, or that the tissues become more sensitive to cytokinins, such that lower levels of the hormone are effective.

Frederick Meins and his colleagues in Basel, Switzerland, have shown that different tissues of tobacco plants exhibit different capacities to synthesize cytokinins in tissue culture. Tobacco pith tissue is cytokinin dependent but capable of being habituated. In contrast, tissues cultured from the cortex of the stem are cytokinin autonomous; they can be cultured indefinitely on medium containing auxin as the only hormone supplement. Finally, tobacco leaf tissue has an absolute requirement for cytokinins and cannot be habituated. These observations indicate that the ability to synthesize cytokinins can be switched on or off during development, and that the on and off states can be stably maintained during cell divisions. Such developmental control of cytokinin synthesis may be important in morphogenesis.

Various Methods are Used to Detect and Identify Cytokinins

Naturally occurring molecules with cytokinin activity may be detected and identified by a combination of physical methods and bioassays. The introduction of immunoaffinity purification methods combined with liquid chromatography and mass spectrometry, along with the use of isotopically labelled cytokinins as internal standards, has resulted in major advances in the measurement of endogenous cytokinins.

Historically, bioassay was the only method available for the identification of molecules that act as cytokinins. Cell proliferation bioassays using tobacco pith tissue or carrot taproot continue to be important because the ability of these tissues to initiate and sustain cell proliferation is proportional to the concentration of active cytokinin in an extract, provided the extract does not contain growth inhibitors.

In addition, some continuously cultured callus tissues of tobacco and soybean that cannot grow in culture without cytokinin have been used for cytokinin bioassays. All of these cytokinin-requiring tissues exhibit a linear increase in growth with increasing cytokinin concentration over a fairly broad range, although high concentrations of cytokinin usually inhibit growth.

Immunological methods are very useful for cytokinin identification and quantification. Researchers can produce antibodies against cytokinins by injecting rabbits or mice with cytokinin ribosides conjugated to a protein. Monoclonal antibodies also have been generated that are highly specific for individual cytokinins. These antibodies can be used to quantitate the amount of a cytokinin in a sample by means of a radioimmunoassay.

For cytokinin isolation, plant extracts are first fractionated, usually by high-performance liquid chromatography (HPLC), and the cytokinins in the fractions are detected and measured by means of a cytokinin radioimmunoassay, similar to the auxin radioimmunoassay.

The cytokinin antibodies can also be used to isolate the hormone from extracts by immunoaffinity chromatography. Immunological methods hold great promise for the identification and quantification of naturally occurring cytokinins because the antibodies are highly specific and more sensitive than most bioassays. Furthermore, these immunological methods are very rapid.

Cytokinins are also Present in Some tRNAs in Animal and Plant Cells

Most organisms contain a variety of modified bases in their tRNA. The normal RNA bases are incorporated into the newly synthesized tRNA chain, and then specific enzyme modify a small subset of the bases. As with the free cytokinins, Ä^2-isopentenyl groups are transferred to the adenine molecules. The enzyme catalyzing this reaction is called tRNA-IPT, which is weakly similar in its sequence to the plant cytokinin biosynthetic enzyme. The similarity was used to help identify the cytokinin biosynthetic enzyme from plants. The tRNA-IPT must be able to recognize a specific base sequence in the tRNA and transfer the isopentenyl group to the adenosine nearest the 3' end of the anticodon. It does not utilize free AMP as a substrate.

The possibility that the free cytokinins are derived from tRNA has been explored extensively. Although the tRNA-bound cytokinins can act as hormonal signals for plant cells if the tRNA is degraded and fed back to the cells, it is unlikely that any significant amount of the free hormonal cytokinin in plants is derived from the turnover of tRNA.

The Ti Plasmid and Plant Genetic Engineering

The Ti plasmid (tumour-inducing plasmid) of *Agrobacterium tumefaciens* has been developed as a vehicle for introducing foreign genes into plants. When *Agrobacterium* infects plants, a region of the Ti plasmid called the T-DNA is taken up by the plant cell and incorporated into one of its chromosomes.

The genes in the T-DNA are referred to as phyto-oncogenes because they induce neoplastic, or tumour-producing, growth (Matzke and Chilton 1981). To use the Ti plasmid as a vector for introducing new genes into plants, it is necessary to disarm the plasmid so that it does not cause tumours. Researchers accomplished this task by deleting the genes in the T-DNA that encode the enzymes controlling auxin and cytokinin synthesis. In addition, it is necessary to introduce a gene into the T-DNA that will enable the investigator to select the transformed cells. Genes for antibiotic resistance are normally used for this purpose. A cloned gene can then be inserted into the T-DNA of the engineered Ti plasmid, and the plasmid can be used to infect cultured cells, leaf discs, or root slices. The infected cells are placed on a culture medium that contains auxin and cytokinin (to induce growth) and the antibiotic. Only the transformed cells can grow in the presence of the antibiotic, because they have received the T-DNA containing not only the foreign gene but also the gene for antibiotic resistance.

To obtain a plant containing the foreign gene, it is necessary to regenerate whole plants from the cultured, transformed cells. Fortunately, methods for accomplishing this regeneration have been developed for many plants, although not for all important crop species yet. The regeneration involves adjusting the ratio of cytokinin to auxin to stimulate both shoot and root formation.

Although the transformation of cereal crops with *Agrobacterium* and their regeneration from cultured cells has been difficult, some remarkable successes have resulted from this approach. Investigators have introduced numerous foreign genes into plants such as tobacco, including soybean storage protein genes. Genes for such desirable characteristics as disease resistance, herbicide resistance, and salt tolerance have been transferred to crop plants by these techniques, and commercial crops now are being grown with these genetically engineered strains. *Arabidopsis* can also be transformed directly by exposing whole plants to a solution containing *Agrobacterium* that is carrying engineered or wild-type Ti plasmids. The plants must be treated in such a way to allow the *Agrobacterium* to enter tissue, either by applying a vacuum or by treating with detergents.

The *Agrobacterium* penetrates the floral tissue and transforms the developing ovules. Isolation of seeds from these *Agrobacterium*-exposed plants yields up to 2% of the seeds that are transformed with the T-DNA. This approach is very useful for molecular genetic studies, such as for characterizing DNA sequences involved in the control of gene expression, or constructing large libraries of insertional mutants.

Cytokinin Can Promote Light-Mediated Development

The *det* mutants of *Arabidopsis* develop many of the characteristics of light-grown seedlings when they are germinated in the dark. The *det* mutants have short, unhooked hypocotyls and expanded cotyledons, and the apical meristem initiates leaves that subsequently expand without light. Although the *det* mutants do not develop full photosynthetic competence in the dark, there is partial development of the chloroplasts, including the expression of genes encoding many photosynthetic enzymes that are not present in the chloroplasts of dark-grown wild-type seedlings.

When wild-type *Arabidopsis* seedlings are germinated in darkness in the presence of cytokinin, they develop many of the characteristics of the *det* mutants: Hypocotyls shorten, cotyledons expand, the apical meristem initiates leaves, and there is partial development of the chloroplasts, including the synthesis of some photosynthetic enzymes. This effect is proportional to the cytokinin dosage given to the seedlings. Cytokinin is said to produce a *phenocopy* of the *det* mutations.

The ability of exogenous cytokinin to cause de-etiolation of dark-grown seedlings is mimicked by certain mutations that lead to cytokinin overproduction. Dark-grown *Arabidopsis* seedlings carrying the *amp1* mutation have zeatin levels that are approximately 2.5 times higher than those of the wild type and develop the characteristics of *det* mutant seedlings. Although the *det* mutants do not have elevated levels of cytokinins, they may have altered sensitivity to cytokinins, since *Arabidopsis* tissues from *det1* plants, when cultured in vitro, do not require cytokinin for growth and become green. Although the functions of the genes altered by the *det* mutants are not completely understood, the *det2* mutation is in a gene required for the synthesis of brassinosteroids. Brassinosteroids and cytokinins may have antagonistic effects in regulating light-induced development. These results suggest that cytokinins may be part of the light signal transduction pathway that is responsible for the initiation of normal vegetative development and photosynthetic competence. However, the role for cytokinins in the developmental pathway initiated by light is not clearly established.

Phylogenetic Tree of Arabidopsis Response Regulators (ARRs)

The top part of the figure shows a phylogenetic tree that represents the degree of relatedness of the receiver domains present in the Arabidopsis genome. The closer two proteins are on the tree, the more similar are their amino acid sequences. Note that these proteins fall into two distinct groups, or clades, called the type-A ARRs and the type-B ARRs.

Localization of Cytokinin Signalling Elements

Expression of *ARR5*. The pattern of *ARR5* expression was examined by fusion of the promoter to a *GUS* reporter gene (A) or by whole-mount *in situ* hybridization (B and C). For the latter, the tissue is hybridized with labelled single-stranded *ARR5* RNA in either the sense orientation (B) or the antisense (C). The sense RNA is a negative control and reveals background, nonspecific staining. The antisense probe specifically hybridizes with the *ARR5* mRNA present in the tissue, thereby revealing its spatial distribution. With both methods, *ARR5* expression is observed primarily in the apical meristems. (From D2 Agostino et al. 2000.)

Ethylene: The Gaseous Hormone

Cloning of the Gene That Encodes ACC Synthase

A major hurdle in the understanding of ethylene biosynthesis was the difficulty in purifying the low abundance, unstable protein and the cloning of the corresponding gene. Researchers overcame this obstacle by raising antisera against partly purified ACC synthase from zucchini fruit slices that were induced to produce large amounts of ethylene. The resulting antisera contained a mixture of antibodies, including antibodies to ACC. The investigators then removed the contaminating antibodies by allowing them to bind to proteins extracted from zucchini tissue that lacked the enzyme. The key to the success of this procedure was the availability of tissue largely lacking the enzyme with which to subtract the non-ACS antibodies from the polyclonal antibody mixture. This circumvented the need for highly purified ACC synthase protein. The purified ACC synthase antibodies were then used to isolate the gene encoding the enzyme.

Once the gene encoding ACS was isolated, it could be used to express the ACC synthase protein in the bacterium *E. coli* at high levels and isolate purified ACC synthase in large amounts. The cloned zucchini gene also allowed genes encoding

ACC synthase to be isolated from numerous other plant species, using low stringency hybridization and degenerate PCR. The deduced amino acid sequences of ACC synthase genes were found to show a high degree of similarity to the *aminotransferase* superfamily of enzymes. These enzymes catalyze the transfer of amino groups among amino acids and require pyridoxal phosphate as a cofactor.

Demonstration of the Dimeric Nature of ACC Synthase

The deduced crystal structure of both apple and tomato ACS reveals a homodimeric protein with shared active sites, which is similar to aminotransferases. The dimeric nature of ACC synthase has been confirmed by an elegant study involving functional complementation. In this study, researchers examined potential dimerization of ACS enzymes by expressing in *E. coli* two ACS proteins that harbour distinct active-site mutations, neither of which has activity on its own.

If ACS acts as a monomer, then expressing the two different mutant versions should not result in an active enzyme, because each is inactive. Conversely, if the two inactive ACS subunits come together as a dimer, the different amino acid changes can complement each other, forming a functional enzyme.

The researchers indeed found that functional ACS enzyme was formed when various pairs of mutants were expressed, which provides compelling evidence that ACS enzymes are able to act as both homo-and heterodimers with shared active sites.

Cloning of the Gene That Encodes ACC Oxidase

The gene that encodes ACC oxidase was cloned by a circuitous route. Researchers isolated a set of genes that are highly induced during tomato fruit ripening. They then blocked the expression of these genes by introducing an antisense version of each into transgenic tomatoes. Blocking of one such gene, *pTOM13*, repressed fruit ripening and inhibited ethylene production. Expression of *pTOM13* in *Xenopus* oocytes as well as in yeast conferred the ability to convert ACC to ethylene, indicating that the gene encodes an ACC oxidase enzyme. Using this gene, researchers then isolated ACC oxidase clones from numerous plant species.

ACC Synthase Gene Expression and Biotechnology

As we have seen, ACC synthase is a cytosolic enzyme that catalyzes the first committed step in ethylene biosynthesis in higher plants. It is a key regulatory enzyme that catalyzes the production of the ethylene precursor ACC from AdoMet (*S*-adenosylmethionine). However, this enzyme is difficult to purify because it is labile and present in low abundance in plant tissues. The molecular cloning and functional expression of ACC synthase in *E. coli* and yeasts have facilitated biochemical and structural studies of this enzyme.

ACC synthase cDNAs and genomic sequences have been cloned from numerous plant species. The emerging picture from the study of these genes is that ACC synthase is encoded by members of a divergent multigene family in which each gene is differentially regulated by various environmental and developmental factors during the growth of a plant. *Arabidopsis*, for example, has twelve and tomato has at least nine gene members, which are differentially regulated by inducers such as auxin, fruit ripening, and wounding.

The protein synthesis inhibitor cycloheximide is a potent inducer of many ACC synthase genes. Cycloheximide inducibility is a characteristic of *primary response genes* such as the early auxin-inducible genes. Two mechanisms have been proposed to explain this inducibility.

- First, the transcription of these genes may be under the control of a shortlived repressor protein whose synthesis is prevented by the addition of cycloheximide;
- Second, the mRNA transcripts of these genes may be shortlived and are stabilized when cycloheximide prevents the synthesis of a nuclease.

The induction of at least one of the genes for ACC synthase (*ACS2*) in *Arabidopsis* by cycloheximide appears to be due to transcriptional activation, and thus is likely regulated by a shortlived transcriptional repressor. A clue to the signal transduction mechanism involved in the regulation of ACC synthase gene expression has come from experiments with lithium (Li^+). Li^+ is known to interfere with the phosphoinositide signalling pathway by inhibiting the activity of inositolphosphate phosphatase. Li^+ is also one of the strongest inducers of ACC synthase activity in plants. This finding suggests that the regulation of at least some ACC synthase genes may involve Ca^{2+} mobilization mediated by inositol trisphosphate (IP_3).

Because ethylene plays such an important role in fruit ripening, the cloning of the genes for ACC synthase and ACC oxidase has enabled scientists to use biotechnology to reduce ethylene production during fruit ripening, and thus to prolong the storage life of fruit.

In the case of tomato, two different strategies have been employed to control ethylene production.

- The first strategy was to use the Ti plasmid of *Agrobacterium tumefaciens* to transform tomato tissue with antisense DNA for ACC synthase or ACC oxidase. When the genes for the two enzymes are expressed in the reverse direction in the cell, the resulting "antisense transcript" binds to and inactivates the "sense transcript" produced by the wild-type gene, effectively shutting down expression of the wild-type gene.

  After transformation, intact plants can be regenerated in tissue culture.
- In the second strategy for reducing ethylene production, tomato tissue was transformed (using *A. tumefaciens*) with genes encoding enzymes that metabolize ACC or AdoMet. In this case, the ACC that is produced is rapidly degraded, and no ethylene can be synthesized.

Regardless of the strategy used, all fruits from these genetically engineered plants show significant delays in ripening. Tomato fruits from plants transformed with the ACC synthase antisense DNA fail to ripen unless treated with ethylene.

Similar approaches are now being carried out in other agronomically important crops, such as melon and banana. In addition, these approaches are being used to retard floral senescence in carnations, thereby reducing postharvest losses.

The research on ACC synthase and other ethylene biosynthesis genes thus provides an excellent example of how basic research can lead to improvements in agricultural and horticultural crops, the full benefits of which we are only beginning to realize.

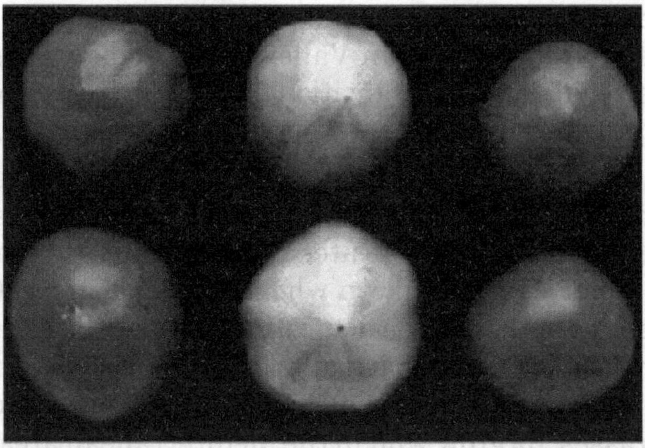

***Figure:** Inhibition of tomato fruit ripening in detached tomato fruits by antisense ACC synthase RNA. Antisense tomato fruits (centre) remain firm and green unless treated by ethylene. Ethylene-treated antisense fruits (left) are indistinguishable from naturally ripened fruits (right) in terms of colour, flavour, aroma, and texture. (Reprinted with permission from Oeller et al. 1991.)*

Abscission and the Dawn of Agriculture

Civilization as we know it is entirely dependent on an abundant food supply, which can be provided only by agriculture. Before the development of agriculture, people lived in small, nomadic groups whose migration patterns followed the seasonal availability of foods derived from wild plants and animals. The shift from hunting and gathering to cultivation and pastoralism led to the establishment of permanent settlements with ever-expanding populations. This transition, considered to be the most significant (and by some, the most disastrous!) cultural change in human history, was made independently at least seven times at various locations around the world.

Remains of some of the earliest settlements are found in the region of the Near East called the Fertile Crescent, a broad arching zone of grassland and woodland that begins at the eastern edge of the Mediterranean and curves around some 2000 km eastward to the Zagros Mountains. One of the oldest known agricultural settlements in the world is found here, in the ancient city of Jericho, located 14 miles north of the modern city of Jericho in the Jordan Valley. Here archaeologists have unearthed evidence for a transition from hunting and gathering to agriculture that occurred around 10,000 years ago.

The first step in the transition from hunting and gathering to agriculture was the domestication of plants and animals. The seven primary domesticates of the Fertile Crescent were barley, emmer wheat, einkorn wheat, sheep, goats, cattle, and pigs. The first of these to be domesticated were the cereals. The Mediterranean region is particularly well stocked with cereals suitable for domestication. Of the world's 56 species of large-seeded grasses, 32 grow wild in the Mediterranean area. Recent DNA analyses have pinpointed the area near the Karacadâg mountains in Turkey as the probable site of origin of einkorn wheat.

All the evidence suggests that wheat and barley were domesticated over a period of only 2 or 3 centuries, between 10,000 and 9700 B.P.. Considering the large number of morphological and physiological differences between even the most primitive of the domesticated forms of wheat and barley today and their wild progenitors, it is difficult to envision how so many changes would come about in such a brief period by natural selection alone. Clearly, some type of artificial selection had to be carried out (most likely by women) to accelerate the evolutionary process. Archaeologists and botanists have attempted to reconstruct this selection process.

Of all the anatomical and physiological differences between wild and domesticated cereals, two are regarded as most crucial. The first is the loss of seed dormancy—the failure of certain seeds to germinate when planted in moist soil unless given additional environmental signals, such as cold or light. The process of plant domestication probably began in small garden plots planted near temporary settlements. It is easy to imagine how the loss of seed dormancy might be selected for by a nomadic group, since seeds that germinated and grew rapidly would be more likely to have been gathered than those whose germination was delayed.

The second critical change that took place during cereal domestication was the development of a tough rachis. The rachis is the main axis of the inflorescence, or spike, of wheat and other cereals, to which the spikelets are attached. Spikelets are small inflorescences bearing one or more florets, or small flowers, along with a set of miniature bractlike leaves. After pollination, the fruits (caryopses) develop within the spikelets.

When the caryopses of the wild-type wheat are fully ripened, a series of abscission layers forms that divides the rachis into dispersal units consisting of a single spikelet attached to a short segment of the rachis. This disarticulation of the rachis begins at the top of the inflorescence and works its way downward. A ripe head of wild wheat is thus easily shattered into dispersal units when touched or blown by the wind and is referred to by archaeologists as a *brittle rachis*.

A brittle rachis is an extremely undesirable trait from a human standpoint. Imagine the plight of a small nomadic group that carefully timed its peregrinations to coincide with the ripening of wild cereals. Because of the extreme fragility of the rachis, it would be crucial to arrive close to the time of ripening, but before disarticulation of the rachis. A slight miscalculation could spell disaster for the group. Arriving too late, they would be greeted by a field of barren stalks yielding only a handful of remnant grains. As a consolation, however, they might find the rare mutant spikes that remained intact because the abscission layers of the rachis failed to develop. This property is referred to by archaeologists as a *tough rachis*. If mutants with the tough-rachis phenotype were then gathered and planted in gardens, their numbers would, over time, increase.

A second scenario for the selection of a tough rachis involves the method of harvesting. The most efficient method for harvesting wheat is to cut it off at the base of the stalk, and evidence suggests that flint-bladed sickles were used for this purpose. However, harvesting wild wheat by cutting would have to be performed before formation of the abscission layers, or the fragile spikes would shatter, sending their spikelets flying to the ground. This method of harvesting would therefore favour the selection of mutations that delayed the abscission process. Hillman and Davies (1990a, 1990b) have demonstrated in field experiments that over many generations, the population would come to be dominated by the varieties with tough rachises. Once cut, the stalks of these varieties of wheat would then be transported to another location, where the grains would be removed by threshing.

Is the mutation rate in wild cereals rapid enough to account for their domestication in a period of only 200 to 300 years? Sharma and Waines (1980) have determined that two genetic loci are involved in the evolution of the tough-rachis character in einkorn wheat. Even if we assume that a mutation that converts a brittle rachis to a tough rachis occurs in only one out of every million wild-type wheat plants, Hillman and Davies (1990a, 1990b) have estimated that varieties with tough rachises would become the dominant forms in wild wheat populations after only 200 years of human management. Higher rates of mutation would shorten this time considerably, possibly to within 20 years.

Thus the experimental results and calculations are in good agreement with the archaeological evidence. Agriculture in the Fertile Crescent, which ultimately led to the founding of Greece and Rome and all of Western civilization, began with mutations in genes that control abscission.

Conservation of Ethylene Signalling Components in Other Plant Species

Genes that are similar to the *Arabidopsis* ethylene signalling components have been found in other plant species. For example, the tomato genome, like that of *Arabidopsis*, encodes a family of ethylene receptor genes, including *Never-ripe*, which are differentially expressed throughout development. The tomato ethylene receptor genes are homologous to ETR1. A tobacco homolog of EIN3 has been found, and genes similar to all the *Arabidopsis* signalling components can be found in the rice genomic sequence. These and other studies suggest that the components of these proteins probably represent ubiquitous mediators of ethylene signalling in plants.

Chapter 6
Plant Pathogens

Plant pathology (also phytopathology) is the scientific study of plant diseases caused by pathogens (infectious diseases) and environmental conditions (physiological factors). Organisms that cause infectious disease include fungi, oomycetes, bacteria, viruses, viroids, virus-like organisms, phytoplasmas, protozoa, nematodes and parasitic plants. Not included are insects, mites, vertebrate or other pests that affect plant health by consumption of plant tissues. Plant pathology also involves the study of pathogen identification, disease etiology, disease cycles, economic impact, plant disease epidemiology, plant disease resistance, how plant diseases affect humans and animals, pathosystem genetics, and management of plant diseases. The *"Disease triangle"* is a central concept of plant pathology. It is based on the principle that infectious diseases develop, or do not develop, based on three-way interactions between the host, the pathogen, and environmental conditions.

The fungi reproduce both sexually and asexually via the production of spores. These spores may be spread long distances by air or water, or they may be soil borne. Many soil borne spores, normally zoospores and capable of living saprotrophically, carrying out the first part of their lifecycle in the soil. Fungal diseases can be controlled through the use of fungicides in agriculture, however new races of fungi often evolve that are resistant to various fungicides. Significant fungal plant pathogens:

Ascomycetes

- o Fusarium spp.
- o Thielaviopsis spp. (Causal agents of: canker rot, black root rot, *Thielaviopsis* root rot)

- o Verticillium spp.
- o Magnaporthe grisea (T.T. Hebert) M.E. Barr; causes blast of rice and gray leaf spot in turfgrasses.

Basidiomycetes

- o Rhizoctonia spp.
- o Phakospora pachyrhizi Sydow; causes Soybean rust
- o *Puccinia* spp.; causal agents of severe rusts of virtually all cereal grains and cultivated gen.

Oomycetes

The oomycetes are not true fungi but are fungal-like organisms. They include some of the most destructive plant pathogens including the genus *Phytophthora* which includes the causal agents of potato late blight and sudden oak death. Despite not being closely related to the fungi, the oomycetes have developed very similar infection strategies and so many plant pathologists group them with fungal pathogens. Significant oomycete plant pathogens:

- Pythium spp.
- Phytophthora spp.; including the causal agent of the Great Irish Famine (1845-1849).

Rice Blast is Hemibiotrophic

Bacteria

Most bacteria that are associated with plants are actually saprotrophic, and do no harm to the plant itself. However, a small number, around 100 species, are able to cause disease. Bacterial diseases are much more prevalent in subtropical and tropical regions of the world. Most plant pathogenic bacteria are rod shaped (bacilli). In order to be able to colonise the plant they have specific pathogenicity factors. Five main types of bacterial pathogenicity factors are known:

1. Cell wall degrading enzymes-used to break down the plant cell wall in order to release the nutrients inside. Used by pathogens such as Erwinia to cause soft rot.
2. Toxins These can be non-host specific, and damage all plants, or host specific and only cause damage on a host plant.

3. Effector Proteins These can be secreted into the extracellular environment or directly into the host cell, often via the Type three secretion system. Some effectors are known to suppress host defence processes.
4. Phytohormones-for example Agrobacterium changes the level of Auxin to cause tumours.
5. Exopolysaccharides-these are produced by bacteria and block xylem vessels, often leading to the death of the plant.

Bacteria control the production of pathogenicity factors via quorum sensing.

Significant bacterial plant pathogens:

- Burkholderia
- Proteobacteria
  - o Xanthomonas spp.
  - o Pseudomonas spp.

Phytoplasmas ('Mycoplasma-like Organisms') and Spiroplasmas

Phytoplasma and Spiroplasma are a genre of bacteria that lack cell walls, and are related to the mycoplasmas which are human pathogens. Together they are referred to as the mollicutes. They also tend to have smaller genomes than true bacteria. They are normally transmitted by sap-sucking insects, being transferred into the plants phloem where it reproduces.

Viruses, Viroids and Virus-like Organisms

There are many types of plant virus, and some are even asymptomatic. Normally plant viruses only cause a loss of yield. Therefore it is not economically viable to try to control them, the exception being when they infect perennial species, such as fruit trees.

Most plant viruses have small, single stranded RNA genomes. These genomes may only encode 3 or 4 proteins: a replicase, a coat protein, a movement protein to allow cell to cell movement though plasmodesmata and sometimes a protein that allows transmission by a vector. Plant viruses must be transmitted from plant to plant by a vector. This is often by an insect (for example, aphids), but some fungi, nematodes and protozoa have been shown to be viral vectors.

Nematodes

Nematodes are small, multicellular wormlike creatures. Many live freely in the soil, but there are some species which parasitize plant roots. They are a problem in tropical and subtropical regions of the world, where they may infect crops. Potato cyst nematodes (Globerodera *pallida* and G. *rostochiensis* are widely distributed in Europe and north and south America and cause $300 million worth of damage in Europe every year. Root knot nematodes have quite a large host range, whereas cyst nematodes tend to only be able to infect a few species. Nematodes are able to cause radical changes in root cells in order to facilitate their lifestyle.

Protozoa

There are a few examples of plant diseases caused by protozoa. They are transmitted as zoospores which are very durable, and may be able to survive in a resting state in the soil for many years. They have also been shown to transmit plant viruses. When the motile zoospores come into contact with a root hair they produce a plasmodium and invade the roots.

Parasitic Plants

Parasitic plants such as mistletoe and dodder are included in the study of phytopathology. Dodder, for example, is used as a conduit for the transmission of viruses or virus-like agents from a host plant to either a plant that is not typically a host or for an agent that is not graft-transmissible.

Physiological Plant Disorders

Significant abiotic disorders can be caused by:

- Natural
- Drought
- Frost damage, and breakage by snow and hail
- Flooding and poor drainage
- Nutrient deficiency
- Salt deposition and other soluble mineral excesses (e.g. gypsum)

- Wind (windburn, and breakage by hurricanes and tornadoes)
- Lightning and wildfire (also often man-made)
- Man-made (arguably not abiotic, but usually regarded as such)
- Soil compaction
- Pollution of air and/or soil
- Salt from winter road salt application
- Herbicide over-application
- Poor education and training of people working with plants (e.g. lawnmower damage to trees)
- Vandalism.

Epidemiology

Management

- Quarantine: Wherein a diseased patch of vegetation or individual plants are isolated from other, healthy growth. Specimens may be destroyed or relocated into a greenhouse for treatment/study. Another option is to avoid introduction of harmful non-native organisms by controlling all human traffic and activity although legislation and enforcement are key in order to ensure lasting effectiveness.
- Cultural: Farming in some societies is kept on a small scale, tended by peoples whose culture includes farming traditions going back to ancient times. (An example of such traditions would be lifelong training in techniques of plot terracing, weather anticipation and response, fertilization, grafting, seed care, and dedicated gardening.) Plants that are intently monitored often benefit not only from active external protection, but a greater overall vigour as well. While primitive in the sense of being the most labour-intensive solution by far, where practical or necessary it is more than adequate.
- Plant resistance: Sophisticated agricultural developments now allow growers to choose from among systematically crossbred species to ensure the greatest hardiness in their crops, as suited for a particular region's

pathological profile. Breeding practices have been perfected over centuries, but with the advent of genetic manipulation even finer control of a crop's immunity traits is possible. (The engineering of foodplants may be less rewarding however, as higher output is frequently offset by popular suspicion and negative opinion about this "tampering" with nature.

- Chemical: Many natural and synthetic compounds exist that could be employed to combat the above threats. This method works by directly eliminating disease-causing organisms or curbing their spread; however it has been shown to have too broad an effect, typically, to be good for the local ecosystem. From an economic standpoint all but the simplest natural additives may disqualify a product from "organic" status, potentially reducing the value of the yield.
- Biological: Crop rotation may be an effective means to prevent a parasitic population from becoming well established, as an organism affecting leaves would be starved when the leafy crop is replaced by a tuberous type, etc. Other means to undermine parasites without attacking them directly may exist.
- Integrated: The use of two or more of these methods in combination offers a higher chance of effectiveness.

How Pathogens Attack Plants

When one considers the relative sizes and masses of host cells/cell walls in light of the necessity of parasites to invade hosts for their own survival, if is obvious that parasites have some mechanism, or mechanisms, for invasion of a suitable host. Mechanical penetration is available to some organisms (certain nematodes and fungi); while others are virtually helpless without chemical or mechanical events.

Cuticle

The cuticle is external to the cell and on the surface of the cell membrane of external cells, usually epidermal cells. The structure of the cuticle is such that though it may show surface topography similar to cellular patterns, it bridges from cell to cell producing a uniform protective barrier to adverse environments of either biotic or abiotic nature. The cuticle is not an amorphous mixture applied to the cell surface but rather is composed many discrete and intermixed components. It is interesting to note there is fact a definite order to the composition of the cuticle. Notice how the components are in layers (cellulose and wax) and in a matrix (lamellae) with cutin providing the interstitial component. The outer surface is primarily, or solely, composed of very hydrophobic molecules (waxes) and the inner surface is very hydrophilic (cellulosic) in composition. There is an admixture of molecules which results in a gradation from hydrophobic to hydrophilic from outer to inner surface (respectively). This provides a means to retard moisture loss from the cell surface while retaining moisture on the cell outer surface.

Appressorium and Penetration Peg

Some elegant research by H. Hoch[R] and colleagues over a number of years has shown that appressoria are extremely sensitive to the topography of leaf surfaces. The growing hypha is able to *sense* very minute alterations in the surface and the spacing between ridges such as occur around stoma. Additionally, there is indication that these mycelia are also chemotactically sensing these stoma. Once *detected* and *discovered* the appressorium is formed over the stomate. Not all appressoria are formed in response to stomatal ridges, nor are all appressoria formed over stoma. Some appressoria form over solid intact cell surfaces.

An appressorium is a highly organized enlarged end of a hypha. Once the hypha *senses* an appropriate site. It enlarges and adheres itself to the leaf surface. This adherence is necessary to support the amount of mechanical force used to penetrate into the plant. When over a stoma, a penetration peg is formed and enters directly through the stoma into the substomatal chamber. From here the fungous hyphae intra-cellularly ramify through the leaf tissue. In most cases, the hyphae grow through the middle lamella between cells extracting their required nutrients without directly killing the host tissue.

Direct penetration is not strictly a mechanical process. Enzymes specific to cuticular components are secreted from the penetration peg as it progresses through the cuticle. It would appear that the cuticular materials server both as inducers of enzymes and well as nutrients; and signals for other metabolic events. Once penetration has occurred a haustorium is often formed from which hyphae emerge and ramify though the leaf tissue in an intra-cellular fashion.

Stylets

Stylets are the nematodes instrument for extracting nutrients from the host. Stylets are exceedingly hard and can conceptually though of a hollow needles. They are characteristic of plant parasites in that all plant parasites have stylets, though not all stylet bearing nemas are plant parasitic. Movement of the Stylet is controlled by muscles which mechanically force the stylet into plant tissue in a probing in and out manner. Nutrients are extracted through the stylet.

Migratory nematodes insert the stylet, feed, remove the stylet and move to another feeding site. Sedentary nematodes attach themselves to a feeding site using the stylet as part of their attachment apparatus.

Plant Disease Management

Plant diseases have caused severe losses to humans in several ways. Starvation and uprooting of families resulted from the Irish famine caused by potato late blight (caused by *Phytophthora infestans*). A valued resource was lost with the virtual elimination of the American chestnut by chestnut blight (caused by *Cryphonectria parasitica*). And direct economic loss such as the estimated one billion dollars lost in one year to American corn growers from southern corn leaf blight (caused by *Cochliobolus maydis*, anamorph *Bipolaris maydis*). Many plant diseases cause less dramatic losses annually throughout the world but collectively constitute sizable losses to farmers and can reduce the aesthetic values of landscape plants and home gardens.

The goal of plant disease management is to reduce the economic and aesthetic damage caused by plant diseases. Traditionally, this has been called plant disease control, but current social and environmental values deem "control" as being absolute and the term too rigid. More multifaceted approaches to disease management, and integrated disease management, have resulted from this shift in attitude, however. Single, often severe, measures, such as pesticide applications, soil fumigation or burning are no longer in common use. Further, disease management procedures are frequently determined by disease forecasting or disease modelling rather than on either a calendar or prescription basis. Disease management might be viewed as proactive whereas disease control is reactive, although it is often difficult to distinquish between the two concepts, especially in the application of specific measures.

This topic is a general overview of some of the many methods, measures, strategies and tactics used in the control or management of plant diseases. Specific management programs for specific diseases are not intended since these will often vary depending on circumstances of the crop, its location, disease severity, regulations and other factors. Most states have some agency such as the Agricultural Extension Service or State Department of Agriculture that formulates and promulgates disease management recommendations for that state. Involvement of these agencies is especially important where the practices include some regulated component such as pesticides or quarantines. Management procedures for some specific crops and diseases can be found in the APSnet Education Centre online plant disease lessons.

Plant disease management practices rely on anticipating occurrence of disease and attacking vulnerable points in the disease cycle (i.e., weak links in the infection chain). Therefore, correct diagnosis of a disease is necessary to identify the pathogen, which is the real target of any disease management program. A thorough understanding of the disease cycle, including climatic and other environmental factors that influence the cycle, and cultural requirements of the host plant, are essential to effective management of any disease.

The many strategies, tactics and techniques used in disease management can be grouped under one or more very broad principles of action. Differences between these principles often are not clear. The simplest system consists of two principles, prevention (prophylaxis in some early writings) and therapy (treatment or cure). The first principle (prevention) includes disease management tactics applied before infection (i.e., the plant is protected from disease), the second principle (therapy or curative action) functions with any measure applied after the plant is infected (i.e., the plant is treated for the disease). An example of the first principle is enforcement of quarantines to prevent introduction of a disease agent (pathogen) into a region where it does not occur. The second principle is illustrated by heat or chemical treatment of vegetative material such as bulbs, corms, and woody cuttings to eliminate fungi, bacteria, nematodes or viruses that are established within the plant material. Chemotherapy is the application of chemicals to an infected or diseased plant that stops (i.e., eradicates) the infection. Although many attempts have been made to utilize chemotherapy, few have been successful. In a few diseases of ornamental or other high value trees, chemotherapy has served as a holding action that must be repeated at intervals of one to several years. For example, antibiotics have been infused into plants to reduce severity of phytoplasma diseases of palms (lethal yellowing) and pears (pear decline) and fungicides have been injected into elms to reduce severity of Dutch elm disease (caused by *Ophiostoma ulmi*) but in all cases the chemotherapeutant must be reapplied periodically. There also are some "systemic" fungicides such as the sterol biosynthesis inhibiting (SBI) and demethylation inhibiting (DMI) fungicides that diffuse into the plant tissues to some extent and eliminate recently established infections.

One early proposal by H. H. Whetzel included four general disease control principles, exclusion, eradication, protection and immunization (the latter principle is more appropriately called resistance since plants do not have an immune system in the same sense as animals). These principles have been expanded or altered to some extent by others. They are still valid and are detailed here but students should investigate other systems such as those proposed by Gäumann, Sharvelle, or the National Academy of Science and use the one(s) that they believe are applicable. These and other disease control principles are discussed in Maloy, Plant Disease Control (1993) cited in the general references of this lesson.

Exclusion

This principle is defined as any measure that prevents the introduction of a disease-causing agent (pathogen) into a region, farm, or planting. The basic strategy assumes that most pathogens can travel only short distances without the aid of some other agent such as humans or other vector, and that natural barriers like oceans, deserts, and mountains create obstacles to their natural spread. In many cases pathogens are moved with their host plants or even on nonhost material such as soil, packing material or shipping containers. Unfortunately, exclusion measures usually only delay the entry of a pathogen, although exclusion may provide time to plan how to manage the pathogen when it ultimately arrives. Karnal bunt (caused by *Tilletia indica*) of wheat is an example of a pathogen originally from India that was anticipated. Measures were established to prevent its introduction, but it finally found its way into the United States. Soybean rust (caused by *Phakopsora pachyrhizi*) has been found recently in the southeastern U.S. and precautions have been undertaken to prevent further spread. Due to its destructiveness, South American leaf blight (SALB) (caused by *Microcyclus ulei*) is a feared disease in the major rubber producing region of Indonesia, and contingency plans have been proposed to chemically defoliate rubber trees by aerial application of herbicides if the pathogen is detected. It is hoped that this would prevent establishment of the pathogen in the region.

In the United States, the Animal and Plant Health Inspection Service (APHIS), a division of the U.S. Department of Agriculture, is responsible for promulgating and enforcing plant quarantine measures. There are also state agencies that deal with local quarantines. Internationally, eight regional plant protection organizations (PPOs) were established in 1951 by the International Plant Protection Convention sponsored by the Food and Agricultural Organization of the United Nations. This was revised in 1997 and now includes nine regional PPOs.

The European and Mediterranean Plant Protection Organization (EPPO) is the oldest of the regional PPOs. The regional PPOs have no regulatory authority such as APHIS or other governmental agency, but function to develop strategies against the introduction and spread of pests and to coordinate the use of phytosanitary regulations to ensure agreement among the different member countries.

An important and practical strategy for excluding pathogens is to produce pathogen-free seed or planting stock through certification programs for seeds and vegetatively propagated plant materials such as potatoes, grapes, tree fruits, etc. These programs utilize technologies that include isolation of production areas, field inspections, and removal of suspect plants to produce and maintain pathogen-free stocks. Planting stock that is freed of pathogens can be increased by tissue culture and micropropagation techniques as well as be maintained in protective enclosures such as screenhouses to exclude pathogens and their vectors. Exclusion may be accomplished by something as simple as cleaning farming equipment to remove contaminated debris and soil that can harbour pathogens such as *Verticillium*, nematodes or other soilborne organisms and prevent their introduction into non-infested fields.

Eradication

This principle aims at eliminating a pathogen after it is introduced into an area but before it has become well established or widely spread. It can be applied to individual plants, seed lots, fields or regions but generally is not effective over large geographic areas. Two large attempts at pathogen eradication in the United States were the golden nematode (*Globodera rostochiensis*) program on Long Island, New York and the citrus canker (caused by *Xanthomonas axonopodis* pv. *citri* and pv. *aurantifolii*) program in Florida. However, neither of these attempts was a lasting success.

Eradication of the golden nematode involved removing infested soil, fumigating soil in infested fields and eventually abandoning infested potato fields for housing developments and other uses. Citrus canker eradication involved widespread removal and burning of diseased trees and, in some cases, destruction of entire citrus groves and nurseries. The disease appeared to be contained and the pathogen eradicated, but the disease has reappeared and new attempts at eradication are ongoing.

Eradication can also be on a more modest scale such as the removal of apple or pear branches infected by the fire blight bacterium (*Erwinia amylovora*) or pruning to remove blister rust cankers (caused by *Cronartium ribicola*) on white pine branches. Or, it can be the sorting and removal of diseased flower bulbs, corms or rhizomes. Hot water seed-treatment of cereal seeds to kill smut mycelium in the seed and heat treatment to eliminate viruses from fruit tree budwood for grafting are other examples of pathogen eradication. Two programs that are actually forms of protection and not pathogen eradication are barberry eradication for reducing stem rust (caused by *Puccinia graminis*) of wheat and *Ribes* eradication for preventing white pine blister rust.

The strategy is that removing these alternate hosts breaks the disease cycles and prevents infection of the economically more valuable host. These two examples are mentioned here because they are frequently cited as eradication measures. However, stem rust can readily spread from wheat to wheat in many regions by the uredinial stage although elimination of the aecial host, barberry, may deter or diminish the development of pathogenic races of the rust.

The white pine blister rust fungus is perennial in the pine host and eradication of the alternate host only protects noninfected trees but does not necessarily eliminate the pathogen from the area.

Eradication may also be accomplished by destroying weeds that are reservoirs of various pathogens or their insect vectors. Elimination of potato cull piles is an effective method of eradicating overwintering inoculum of the late blight pathogen.

Soil fumigation has been a widely used eradication strategy. This technology involves introducing gas-forming chemicals such as carbon disulfide, methyl bromide, or chloropicrin into soil to kill target pathogens. However, undesirable side effects such as killing beneficial organisms, contamination of groundwater, and toxicity of these chemicals have resulted in less reliance on this approach for disease management.

Volatile fumigants like methyl bromide are injected into soil and sealed with a plastic film. Some water-soluble fumigants like metam sodium can be injected into the soil and the soil simply compacted to form a seal.

Crop rotation is a frequently used strategy to reduce the quantity of a pathogen, usually soilborne organisms, in a cropping area. Take-all of wheat (caused by *Gaeumannomyces graminis*) and soybean cyst nematode (*Heterodera glycines*) are two examples of soilborne diseases that are easily managed by short rotations of 1 and 2 years, respectively, out of susceptible crops, which may include susceptible weed hosts such as grasses in the case of take-all.

Burning is an effective means of eradicating pathogens and is often required by law to dispose of diseased elm trees affected by Dutch elm disease (DED), citrus trees infected by citrus canker or of bean fields infected by halo blight bacteria (*Pseudomonas syringae* pv. *phaseolicola*). Propane flaming can effectively destroy *Verticillium* microsclerotia in mint stems, and flaming potato stems prior to harvest may prevent tuber infection by the late blight pathogen. However, burning agricultural fields is controversial because the smoke creates human health and safety and environmental concerns.

Students may note that the principle of eradication is a good example of the conflicting concepts of some of these general principles for at least two reasons. One is that some of the examples above could be placed as readily under protection as under eradication. The second is that complete eradication of pathogens, especially from large areas is rarely accomplished.

Protection

This principle depends on establishing a barrier between the pathogen and the host plant or the susceptible part of the host plant. It is usually thought of as a chemical barrier, e.g., a fungicide, bactericide or nematicide, but it can also be a physical, spatial, or temporal barrier. The specific strategies employed assume that pathogens are present and that infection will occur without the intervention of protective measures. For example, bananas are covered with plastic sleeves as soon as the fruit are set to protect the fruit from various pests including fruit decay fungi.

Protection often involves some cultural practice that modifies the environment, such as tillage, drainage, irrigation, or altering soil pH. It may also involve changing date or depth of seeding, plant spacing, pruning and thinning, or other practices that allow plants to escape infection or reduce severity of disease. Raising planting beds to assure good soil water drainage is an example of cultural management of plant diseases such as root and stem rots. Fungicides have been used for more than a hundred years and new fungicides continue to be developed. Bordeaux mixture, a basic copper sulfate fungicide, was the first widely used fungicide and is still used today in various forms.

The earliest fungicides were simple elements like sulfur or metallic compounds of copper or mercury, and these are generally classed as inorganic fungicides. In the early to mid-1900s organic fungicides such as thiram, captan, and the bisdithiocarbamates were developed. These are broad-spectrum, contact or protectant fungicides that control a wide range of fungal diseases.

Starting in the 1960s the "systemic" fungicides were developed. Most of these are not truly systemic in plants but have some limited mobility, usually translaminar, and often give some post-infection benefits. Some of the "systemic" fungicides move upward in the plant's vascular system, but currently only one (fosetyl-Al) has ambimobile distribution (both upward and downward) that would constitute a truly systemic fungicide. In addition to the SBI and DMI fungicides mentioned earlier, a recent group of systemic fungicides are the strobilurins fungicides: Benefits and risks) Some fungicides have narrow ranges of activity and are used primarily for control of specific groups of diseases such as downy mildews, rusts, smuts or powdery mildews while others are active against a wider range of diseases.

One liability of these recent narrow-range fungicides is that they often have single-site modes of action, (that is, their site-specific activity is controlled by one or a few genes), and thus are especially prone to development of fungicide resistance in the pathogen. Several management strategies have been developed to combat fungicide resistance. These include using mixtures of single-site and multi-site fungicides, alternating applications of fungicides with different modes of action, applying fungicides only when needed instead of on either a calendar or prescription basis, and applying the recommended dosage and not attempting to cut costs by reducing the recommended amount of fungicide applied.

Fungicides can be applied by any of several methods: ground sprayers, airplanes or through irrigation systems, but to be effective applications must be done properly. First, the fungicide must be legally registered for use on the plant involved and against the target disease. Several different chemicals may be registered for the same crop or disease.

If the different fungicides are similar in effectiveness, cost, ease of application, and safety, then timing of application becomes the most critical factor. If applied too early much of the chemical will be wasted before it can be effective; if applied too late, it will be largely ineffective. The benefits of properly applied fungicides can often be striking. Distribution of the spray droplets is important; the finer the spray the more complete the coverage on the plant surface. However, very small droplets form a mist that is easily displaced by wind.

Many cultural practices can be modified to manage the occurrence, intensity or severity of plant diseases. These include selection of suitable growing sites for the crop, adequate tillage to bury pathogen-infested plant residues, rotation to nonsusceptible crops, selecting pathogen-free planting stocks, orientation of plantings to improve exposure to sun and air currents, pruning and thinning to eliminate sources of infection and improve aeration in and around susceptible plants, water management on both plants and in soil, adequate nutrition, proper cultivation to improve root growth and avoid plant injury, and sanitation procedures to eliminate sources of inoculum.

Biological control involves the use of one living organism to control another, and this management technology has received much attention in recent times. However, the number of biological agents registered for use is relatively small, success has been limited, and application has been largely restricted to intensively managed, high value crops such as greenhouse plants.

Two examples of effective biological control are the use of the fungus *Peniophora gigantea* to inoculate tree stumps to prevent infection of adjacent trees by the wood decay fungus *Heterobasidion annosum*, and the application of the nonpathogenic (i.e., non-tumour-producing) bacterium *Agrobacterium radiobacter* to fruit trees before planting to prevent infection by the crown gall bacterium (*Agrobacterium tumefaciens*).

Resistance

Use of disease-resistant plants is the ideal method to manage plant diseases, if plants of satisfactory quality and adapted to the growing region with adequate levels of durable resistance are available. The use of disease-resistant plants eliminates the need for additional efforts to reduce disease losses unless other diseases are additionally present. Resistant plants are usually derived by standard breeding procedures of selection and/or hybridization. A few disease-resistant lines have been obtained by inducing mutations with x-rays or chemicals. There is also interest in chemicals called "plant activators" that induce plant defence responses called systemic acquired resistance (SAR) and induced resistance. Recently, resistant plants have been developed through the use of genetic engineering.

Selection of resistant plants involves subjecting plants to high levels of disease pressure and using the surviving plants as sources of disease resistance. Plants that survive this pressure often have genetic resistance that can be utilized directly by propagation or as sources of resistance to develop resistant plants that also have the requisite qualities for that crop. Hybridization is a tactic where a plant having the desired agronomic or horticultural qualities, but is susceptible to a disease, is crossed with a plant that is resistant but which may or may not have the other desirable characteristics such as size, yield, flavour, aesthetics, etc.

Disease escape occurs when susceptible plants do not become diseased for some reason. This may be due to some anatomical or physical character, such as the occurrence of leaf hairs, thick cuticle, or modified stomata, or they may be environmental, in which conditions are not conducive to disease development. Although disease escape based on some anatomical feature is useful occasionally, escape more often complicates the process of developing disease resistant plants. Development of disease-resistant plants has been relatively successful with annual and biennial plants, but less so with perennials, primarily because of the longer time required to develop and test the progeny. Woody perennials, such as ornamental, forest, and orchard trees, have been especially difficult for plant breeders to develop useful disease resistance.

For example, chestnut blight and Dutch elm disease have devastated two valued native trees. In both cases there have been extensive attempts to develop resistant trees, usually by creating hybrids with exotic chestnut or elm trees, and some resistant selections have resulted. Unfortunately, these generally lack the desirable qualities, such as nut flavour or tree forms characteristic of the native trees. Another introduced disease that has impacted native trees is white pine blister rust. There has been an intense effort for more than 50 years to select and improve rust-resistant pines from the surviving population. These trees are now being planted for reforestation, but it will be another 50 or so years, when these trees have matured to produce a timber crop, before the success of this program is known. Development of resistance has been most successful against the more specialized pathogens such as rust fungi, smut fungi, powdery mildew fungi, and viruses, but less so against general pathogens such as many blight, canker, root rot and leaf spotting pathogens.

A major problem with genetically resistant plants is that host-differentiated pathogenic races can be selected, so that many breeding programs become continuous processes to develop disease resistant plant lines. Disease resistance conferred by a single major gene is sometimes called specific or qualitative resistance and is race-specific. This type of resistance is often unstable, and emergence of a pathogenic race that can attack that genotype can completely overcome this type of resistance. Quantitative resistance or general resistance derives from many different genes for resistance with additive effects to provide more stable (or durable) resistance to pathogens.

There are several strategies to minimize this race development and resistance failure. These include methods of gene deployment, where different genetic plant types are interspersed on a regional basis to avoid a genetic monoculture, or planting mixtures of cultivars having different genetic compositions to ensure that some component of the crop will be resistant to the disease.

A recent and controversial technique in developing disease resistant plants is the insertion of genes from other organisms into plants to impart some characteristic. For example, genes from the bacterium *Bacillus thuringiensis* have been inserted into plants to protect against insect attacks. Plants with these inserted genes are called genetically-modified organisms (GMOs), and have caused concern that unanticipated, and perhaps detrimental, characteristics, such as unforeseen allergens, may also be transferred to the new plants. However, unforeseen and undesirable qualities also can be transmitted by conventional plant breeding techniques.

The potato cultivar Lenape was developed in part because of its resistance to *Potato virus A* and resistance to late blight tuber infection. After it was released it was discovered that the tubers contained very high levels of solanine, a toxic alkaloid. The wheat cultivar Paha had resistance to stripe rust (caused by *Puccinia striiformis*) but also was very susceptible to flag smut (caused by *Urocystis agropyri*). Both of these plant cultivars, developed by conventional breeding methods, were quickly taken out of production. There is much interest in the genetic engineering of disease-resistant plants and some success has been obtained with several virus diseases, the best known of which is papaya ringspot. This approach to plant disease management will likely expand, especially for widely grown crops such as wheat, corn, soybeans, rice, and the like, as social, legal, and economic obstacles are overcome.

Integrated Disease Management

Integrated Disease Management (IDM) is a concept derived from the successful Integrated Pest Management (IPM) systems developed by entomologists for insect and mite control. In most cases IDM consists of scouting with timely application of a combination of strategies and tactics. These may include site selection and preparation, utilizing resistant cultivars, altering planting practices, modifying the environment by drainage, irrigation, pruning, thinning, shading, etc., and applying pesticides, if necessary. But in addition to these traditional measures, monitoring environmental factors (temperature, moisture, soil pH, nutrients, etc.), disease forecasting, and establishing economic thresholds are important to the management scheme. These measures should be applied in a coordinated integrated and harmonized manner to maximize the benefits of each component. For example, balancing fertilizer applications with irrigation practices helps promote healthy vigorous plants. However, this is not always easy to accomplish, and "disease management" may be reduced to single measures exactly the same as the ones previously called "disease control." Whatever the measures used, they must be compatible with the cultural practices essential for the crop being managed.

Plant Diseases

Like human beings and other animals, plants are subject to diseases. In order to maintain a sufficient food supply for the world's population, it is necessary for those involved in plant growth and management to find ways to combat plant diseases that are capable of destroying crops on a large scale. There are many branches of science that participate in the control of plant diseases. Among them are biochemistry, biotechnology, soil science, genetics and plant breeding, meteorology, mycology (fungi), nematology (nematodes), virology (viruses), and weed science. Chemistry, physics, and statistics also play a role in the scientific maintenance of plant health. The study of plant diseases is called plant pathology.

The most common diseases of cultivated plants are bacterial wilt, chestnut blight, potato late blight, rice blast, coffee rust, stem rust, downy mildew, ergot, root knot, and tobacco mosaic. This is a small list of the more than 50,000 diseases that attack plants. Diseases can be categorized as annihilating, devastating, limiting, or debilitating. As the term suggests, annihilating diseases can totally wipe out a crop, whereas a devastating plant disease may be severe for a time and then subside. Debilitating diseases weaken crops when they attack them successively over time and limiting diseases reduce the viability of growing the target crop, thereby reducing its economic value. Plant diseases are identified by both common and scientific names. The scientific name identifies both the genus and the species of the disease-causing agent.

For the past 50 years, the ability to combat plant diseases through the use of modern farm management methods, fertilization of crops, irrigation techniques, and pest control have made it possible for the United States to produce enough food to feed its population and to have surpluses for export. However, the use of pesticides, fungicides, herbicides, fertilizers and other chemicals to control plant diseases and increase crop yields also poses significant environmental risks. Air, water, and soil can become saturated with chemicals that can be harmful to human and ecosystem health.

History of Plant Pathology

While early civilizations were well aware that plants were attacked by diseases, it was not until the invention of the first microscope that people began to understand the real causes of these diseases. There are references in the Bible to blights, blasts, and mildews. Aristotle wrote about plant diseases in 350 B.C. and Theophrastus (372-287 B.C.) theorized about cereal and other plant diseases. During the Middle Ages in Europe, ergot fungus infected grain and Shakespeare mentions wheat mildew in one of his plays.

After Anton von Leeuwenhoek constructed a microscope in 1683, he was able to view organisms, including protozoa and bacteria, not visible to the naked eye. In the eighteenth century, Duhumel de Monceau described a fungus disease and demonstrated that it could be passed from plant to plant, but his discovery was largely ignored. About this same time, nematodes were described by several English scientists and by 1755 the treatment of seeds to prevent a wheat disease was known.

In the nineteenth century, Ireland suffered a devastating potato famine due to a fungus that caused late blight of potatoes. At this time, scientists began to take a closer look at plant diseases. Heinrich Anton DeBary, known as the father of modern plant pathology, published a book identifying fungi as the cause of a variety of plant diseases. Until this time, it was commonly believed that plant diseases arose spontaneously from decay and that the fungi were caused by this spontaneously generated disease. DeBary supplanted this theory of spontaneously generated diseases with the germ theory of disease. Throughout the rest of the nineteenth century scientists working in many different countries, including Julian Gotthelf Kühn, Oscar Brefeld, Robert Hartig, Thomas J. Burrill, Robert Koch, Louis Pasteur, R. J. Petri, Pierre Millardet, Erwin F. Smith, Adolph Mayer, Dimitri Ivanovski, Martinus Beijerinck, and Hatsuzo Hashimoto, made important discoveries about specific diseases that attacked targeted crops. During the twentieth century advances were made in the study of nematodes. In 1935 W. M. Stanley was awarded a Nobel Prize for his work with the tobacco mosaic virus. By 1939, virus particles could be seen under the new electron microscope. In the 1940s fungicides were developed and in the 1950s nematicides were produced. In the 1960s Japanese scientist Y. Doi discovered mycoplasmas, organisms that resemble bacteria but lack a rigid cell wall, and in 1971, T. O. Diener discovered viroids, organisms smaller than viruses.

Causes of Plant Disease

Plant diseases can be infectious (transmitted from plant to plant) or noninfectious. Noninfectious diseases are usually referred to as disorders. Common plant disorders are caused by deficiencies in plant nutrients, by waterlogged or polluted soil, and by polluted air. Too little (or too much) water or improper nutrition can cause plants to grow poorly. Plants can also be stressed by weather that is too hot or too cold, by too little or too much light, and by heavy winds. Pollution from automobiles and industry, and the excessive application of herbicides (for weed control) can also cause noninfectious plant disorders. Infectious plant diseases are caused by pathogens, living microorganisms that infect a plant and deprive it of nutrients. Bacteria, fungi, nematodes, mycoplasmas, viruses and viroids are the living agents that cause plant diseases. Nematodes are the largest of these agents, while viruses and viroids are the smallest. None of these pathogens are visible to the naked eye, but the diseases they cause can be detected by the symptoms of wilting, yellowing, stunting, and abnormal growth patterns.

Bacteria

Some plant diseases are caused by rod-shaped bacteria. The bacteria enter the plant through natural openings, like the stomata of the leaves, or through wounds in the plant tissue. Once inside, the bacteria plug up the plant's vascular system (the vessels that carry water and nutrients) and cause the plant to wilt. Other common symptoms of bacterial disease include rotting and swollen plant tissues. Bacteria can be spread by water, insects, infected soil, or contaminated tools. Bacterial wilt attacks many vegetables including corn and tomatoes, and flowers. Crown gall, another bacterial plant disease, weakens and stunts plants in the rose family and other flowers. Fireblight attacks apple, pear, and many other ornamental and shade trees.

Fungi

About 80% of plant diseases can be traced to fungi, which have a great capacity to reproduce themselves both sexually and asexually. Fungi can grow on living or dead plant tissue and can survive in a dormant stage until conditions become favourable for their proliferation. They can penetrate plant tissue or grow on the plant's surface. Fungal spores, which act like seeds, are spread by wind, water, soil, and animals to other plants. Warm, humid conditions promote fungal growth. While many fungi play useful roles in plant growth, especially by forming mycorrhizal associations with the plant's roots, others cause such common plant diseases as anthracnose, late blight, apple scab, club root, black spot, damping off, and powdery mildew. Many fungi can attack are variety of plants, but some are specific to particular plants.

The list of fungi and the plants they infect is a long one. Black spot attacks roses, while brown rot damages stone fruits. Damping off is harmful to seeds and young plants. Downy mildew attacks flowers, some fruits, and most vegetables. Gray mold begins on plant debris and then moves on to attack flowers, fruits, and vegetables. Oak root fungus and oak wilt are particularly damaging to oaks and fruit trees. Peach leaf curl targets peaches and nectarines. Powdery mildew, rust, sooty mold, and southern blight attack a wide variety of plants, including grasses. Texas root rot and water mold root rot can also infect many different plants. Verticillium wilt targets tomatoes, potatoes, and strawberries.

Viruses and Viroids

The viruses and viroids that attack plants are the hardest pathogens to control. Destroying the infected plants is usually the best control method, since chemicals to inactivate plant viruses and viroids have not proven effective. While more than 300 plant viruses have been identified, new strains continually appear because these organisms are capable of mutating.

The symptoms of viral infection include yellowing, stunted growth in some part of the plant, and plant malformations like leaf rolls and uncharacteristically narrow leaf growth. The mosaic viruses can infect many plants. Plants infected with this virus have mottled or streaked leaves; infected fruit trees produce poor fruit and a small yield.

Nematodes

Nematodes are tiny microscopic animals with wormlike bodies and long, needlelike structures called stylets that suck nutrients from plant cells. They lay eggs that hatch as larvae and go through four stages before becoming adults. Nematodes have a 30-day life cycle, but they can remain in a dormant state for more than 30 years. Nematicides are chemicals used to control nematode infestations. Marigolds are resistant to nematodes and are often planted to help eliminate them from infected soil.

Nematodes primarily attack plant roots, but they may also destroy other parts of the plant either internally or externally. They thrive in warm, sandy, moist soil and attack a variety of plants including corn, lettuce, potatoes, tomatoes, alfalfa, rye, and onions. However, all nematodes are not harmful to plants. Some are actually used to control other plant pests such as cutworms, armyworms, and beetle grubs.

Other Causes of Plant Diseases

Mycoplasmas are single-celled organisms that lack rigid cell walls and are contained within layered cell membranes. They are responsible for the group of plant diseases called yellow diseases and are spread by insects such as the leafhopper. Parasitic plants, such as mistletoe, cannot get their nutrients from the soil, but must attach themselves to other plants and use nutrients from the host plant to survive. They weaken the wood of their host trees and deform the branches.

Disease Cycles

An equilateral disease triangle is often used to illustrate the conditions required for plant diseases to occur. The base of the triangle is the host and the two equal sides represent the environment and the pathogen. When all three factors combine, then disease can occur. Pathogens need plants in order to grow because they cannot produce their own nutrients. When a plant is vulnerable to a pathogen and the environmental conditions are right, the pathogen can infect the plant causing it to become diseased. Plant disease control is achieved by changing the host plant, by destroying the pathogen or by changing the plant's environment. The key to success in growing plants, whether in the home garden or commercially, is to change one or more of the three factors necessary to produce disease. Disease-resistant plants and enrichment of soil nutrients are two ways of altering the disease triangle.

Weather is one environmental factor in the plant disease triangle that is impossible to control. When weather conditions favour the pathogen and the plant is susceptible to the pathogen, disease can occur. Weather forecasting provides some help; satellites monitor weather patterns and provide farmers with some advance warning when conditions favourable to disease development are likely to occur. Battery-powered microcomputers and microenvironmental monitors are place in orchards or fields to monitor temperature, rainfall, light levels, wind, and humidity. These monitors provide farmers with information that helps them determine the measures they need to take to reduce crop loss due to disease.

Control

Control of plant disease begins with good soil management. The best soil for most plants is loamy, with good drainage and aeration. This minimizes diseases that attack the roots and allows the roots to feed nutrients from the soil to the rest of the plant. Organic methods, such as the addition of compost, can improve soil quality, and fertilizers can be added to the soil to enrich the nutrient base. Soil pH measures the degree of acidity or alkalinity of the soil. Gardeners and farmers must be aware of the pH needs of their plants, since the right pH balance can help reduce susceptibility to disease, especially root diseases like club root or black root rot.

Other important factors in the control of plant disease are the selection of disease-resistant plants (cultivars), proper watering, protection of plants from extreme weather conditions, and rotation of crops. Disposal of infected plants is important in the control of diseases, as is the careful maintenance of tools and equipment used in farming and gardening. Many plant diseases can easily be spread by hand and by contact with infected tools, as well as by wind, rain, and soil contamination. Plant diseases can also be spread by seeds, and by transplants and cuttings; careful attention to the presence of disease in seeds, transplants, and cuttings can avoid the spread of pathogens. Crop rotation is an important part of reducing plant diseases. Pathogens that favour a specific crop are deprived of their preferred host when crops are rotated. This reduces the virulence of the pathogen and is a natural way to reduce plant disease. Soil solarization is another natural method used by gardeners to reduce diseases.

Barriers or chemical applications to eliminate pests that may carry pathogens to plants are another method of disease control. The use of chemical pesticides has become standard practice among home gardeners and commercial growers alike. Among the organic chemicals used today are copper, lime-sulfur, Bordeaux mixture, fungicidal soap, and sulfur. After World War II, DDT, a synthetic insecticide, was used to destroy plant pests. Today, the use of this and a number of other pesticides has been banned or restricted because they were found to present hazards to the health of human, wildlife, and the environment.

Plant Breeding

Plant breeding began when early humans saved seeds and planted them. The cultural change from living as nomadic hunter-gatherers, to living in more settled communities, depended on the ability to cultivate plants for food. Present knowledge indicates that this transition occurred in several different parts of the world, about 10,000 years ago. Today, there are literally thousands of different cultivated varieties (cultivars) of individual species of crop plants. As examples, there are more than 4,000 different peas (*Pisum sativum*), and more than 5,000 grape cultivars, adapted to a wide variety of soils and climates.

The methods by which this diversity of crops was achieved were little changed for many centuries, basically requiring observation, selection, and cultivation. However, for the past three centuries most new varieties have been generated by deliberate cross-pollination, followed by observation and further selection.

The science of genetics has provided a great deal of information to guide breeding possibilities and directions. Most recently, the potential for plant breeding has advanced significantly, with the advent of methods for the incorporation of genes from other organisms into plants via recombinant DNA-techniques. This capacity is broadly termed "genetic engineering." These new techniques and their implications have given rise to commercial and ethical controversies about "ownership," which have not yet been resolved.

Somaclonal Variation

Replicate plant cells or protoplasts that are placed under identical conditions of tissue culture do not always grow and differentiate to produce identical progeny (clones). Frequently, the genetic material becomes destabilized and reorganized, so that previously-concealed characters are expressed. In this way, the tissue-culture process has been used to develop varieties of sugar cane, maize, rapeseed, alfalfa, and tomato that are resistant to the toxins produced by a range of parasitic fungi. This process can be used repeatedly to generate plants with multiple disease resistance, combined with other desirable characters.

Vectors for Gene Transfer

Agrobacterium tumefaciens and *A. rhizogenes* are soil bacteria that infect plant roots, causing crown gall or "hairy roots" diseases. Advantage has been taken of the natural ability of *Agrobacterium* to transfer plasmid DNA into the nuclei of susceptible plant cells. *Agrobacterium* cells with a genetically-modified plasmid, containing a gene for the desired trait and a marker gene, usually conferring antibiotic resistance, are incubated with protoplasts or small pieces of plant tissue. Plant cells that have been transformed by the plasmid can be selected on media containing the antibiotic, and then cultured to generate new, transgenic plants. Many plant species have been transformed by this procedure, which is most useful for dicotyledonous plants. The gene encoding Bt, as well as genes conferring resistance to viral diseases, have been introduced into plants by this method.

Direct Gene Transfer

Two methods have been developed for direct gene transfer into plant cells—electroporation and biolistics. Electroporation involves the use of high-voltage electric pulses to induce pore formation in the membranes of plant protoplasts. Pieces of DNA may enter through these temporary pores, and sometimes protoplasts will be transformed as the new DNA is stably incorporated. New plants are then derived from cultured protoplasts. This method has proven valuable for maize, rice, and sugar cane, species that are outside the host range for vector transfer by *Agrobacterium*.

Biolistics refers to the bombardment of plant tissues with microprojectiles of tungsten coated with the DNA intended for transfer. Surprisingly, this works. The size of the particles and the entry velocity must be optimized for each tissue, but avoiding the need to isolate protoplasts increases the potential for regenerating transformed plants. Species that cannot yet be regenerated from protoplasts are clear candidates for transformation by this method.

Genetically-modified Plants

In 1992, a tomato with delayed ripening became the first genetically-modified (GM) commercial food crop. More than 40 different GM crops are now being grown commercially. GM corn and cotton contain bacterial genes that kill insects and confer herbicide-resistance on the crops. GM squash contains viral genes that confer resistance to viruses.

Potatoes carry the Bt gene to kill the Colorado potato beetle and a viral gene that protects the potato from a virus spread by aphids. Mauve-coloured carnations carry a petunia gene required for making blue pigment. In many cases, GM crops result in increased yields and reduced use of pesticides. New research is focused on producing GM foods containing increased vitamins and human or animal vaccines.

GM crops are very controversial. There is concern that the widespread dissemination of the Bt gene will cause insects to become resistant. It has been reported that pollen from Bt corn is toxic to the caterpillars of monarch butterflies. It also is possible that GM crops will interbreed with wild plants, resulting in “superweeds” resistant to herbicides. There is also concern that the antibiotic-resistance genes, used as markers for gene transfer, may be passed from the plants to soil microorganisms or bacteria in humans who eat the food. Finally, the possibility of allergic reactions to the new compounds in food exists. Many countries have banned the production and importation of GM crops.

History of Plant Pathology

The study of plant pathology in New Zealand was begun in 1893 with the appointment of T. W. Kirk as Biologist to the newly formed Department of Agriculture. Apart from undertaking a considerable amount of work on the identification of diseases, Kirk did much to educate farmers and fruitgrowers to appreciate what diseases were and how they could best be controlled.

In 1911 when Kirk became Director of the Horticulture Division, A. H. Cockayne was placed in charge of the Biological Laboratory, which was moved from Wellington in 1915 to the Department's Central Development Farm at Weraroa, near Levin. The development of the Laboratory was restricted by the First World War. The post-war years saw the beginning of a new era in plant pathology in this country. G. H. Cunningham, who joined the staff in 1919 as mycologist, began a systematic survey of diseases attacking fruits, field crops, vegetables, and flowers. His work on the classification of fungi was also begun about this time. In 1925 he published the first New Zealand book on plant diseases – *Fungous Diseases of Fruit Trees in New Zealand.*

The Biological Laboratory returned in 1920 to Wellington where it remained until 1928 when the staff was transferred to Palmerston North to form the nucleus of the Plant Research Station, an organisation administered jointly by the Departments of Agriculture and of Scientific and Industrial Research. Within the new Station, Cunningham was placed in charge of a Mycological Laboratory. The next eight years was a period of considerable development and expansion in plant pathology. Investigations which until then had been mainly concerned with fungous and bacterial diseases were widened in scope to include virus and physiological diseases, techniques of spray application, and evaluation of therapeutants used for disease control.

In 1936, following a reorganisation of agricultural research in New Zealand, the Department of Scientific and Industrial Research was made responsible for plant research and the Plant Research Station disbanded. As a result of the reorganisation, the Plant Diseases Division was established as a branch of the Department of Scientific and Industrial Research with G. H. Cunningham as Director. The Division moved in 1939 to permanent quarters at Mount Albert, Auckland, where modern laboratory and glasshouse facilities had been provided on an area of 16½ acres.

Plant Diseases Division, the centre of plant pathological work, has a permanent staff of 70 including 24 professional plant pathologists. Of the latter, 19 are stationed at Mount Albert where they are grouped in teams working in the fields of systematic mycology, economic mycology, bacteriology, virology, and therapeutant control of diseases, each group being responsible for both basic and applied research in its own specialist field. The remainder are stationed at substations, disease problems of field crops being studied by three professional officers at Lincoln in the South Island, while diseases of pasture plants are investigated by an officer at Palmerston North, and tobacco diseases by one at Nelson.

In addition to those employed by the Plant Diseases Division, there are about 12 other plant pathologists practising their profession. Within the Department of Scientific and Industrial Research, the Crop Research Division employs a pathologist to investigate virus diseases of potatoes, the Fruit Research Division takes part in the investigation of fruit tree viruses, and a nematologist is attached to the Entomology Division. To assist with plant quarantine and advisory and extension work, the Department of Agriculture also employs several plant pathologists. In addition to their servicing work, these officers undertake research on diseases of vegetables and berry fruits and on chemical control of diseases. Research on diseases of forest trees is carried out by the plant pathologists employed by the Forest Research Institute of the State Forest Service. At Massey and Lincoln Agricultural Colleges, students are trained by qualified scientists who also undertake investigation of plant diseases. At least one plant pathologist is employed by a chemical-manufacturing firm.

Scientists in other fields also contribute to work in plant pathology. Thus plant breeders at the Crop Research and Grasslands Divisions of the Department of Scientific and Industrial Research are concerned with the breeding of disease-resistant varieties of field crop, vegetable, and pasture plants, and chemists of the Dominion Laboratory assist with studies on therapeutant chemicals.

An extension and advisory service concerning plant diseases is provided by the Farm Advisory and Horticultural Divisions of the Department of Agriculture, whose officers make substantial contributions to plant pathology with their field observations and by carrying out plant-disease surveys and conducting field trials on disease control. They also assist pathologists by drawing to their attention new disease problems, collecting disease material for identification and study, and by assisting in the selection of disease-resistant varieties. The Department of Agriculture also provides a plant quarantine service and administers seed certification schemes which have as one of their aims the production and distribution of disease-free seed. Introduction and use of control measures have been successful in greatly reducing losses caused by plant diseases, but in spite of this and the fact that hundreds of thousands of pounds are spent annually on disease control, direct losses greatly exceed £10,000,000 and are probably considerably in excess of £20,000,000 a year. Also of major importance are indirect losses brought about by the inability, because of disease, to grow economically certain crops, such as brassicas and cereals, in many parts of the country.

At a moderate estimate the combined ravages of fungous, bacterial, and virus diseases reduce the value of our brassica crops, which are widely used as supplementary stock food, by £5,000,000 a year, while diseases of fruit crops take an annual toll of more than £1,000,000. In some seasons losses in cereal crops also amount to more than £1,000,000, and diseases in vegetables regularly cause losses of the order of millions of pounds a year. No crop grown in the Dominion is unaffected.

Defence Against Plant Diseases

The main concern of plant pathologists in New Zealand is to gain an understanding of our diseases and their significance in the country's economy and to minimise the losses they cause, generally in one or more of the following three ways: (1) By preventing the introduction of diseases not already here; (2) By eradicating those not fully established; and (3) By control of those which cause damage with chemical or cultural methods or by the development of resistant varieties.

Some serious plant diseases now confined to other countries, if established here, could cause losses amounting to millions of pounds annually. Hence plant quarantine regulations are enforced. Plants and seeds not known to carry diseases or pests which could be dangerous to our crops may be imported without restriction but are inspected at the port of entry. Others require an entry permit, must be grown for one season in a specified area and kept under observation by officers of the Department of Agriculture who certify them as disease-free before distribution. The introduction of plants of a third category is prohibited except in small numbers under special licence. These are grown under strict quarantine conditions at the Plant Diseases Division, Mount Albert, and released only after they have been declared free from disease.

Although eradication has been attempted on a number of occasions, it is rarely practicable because a disease is usually well established by the time it is recognised. An attempt to eradicate fire blight (*Erwinia amylovora*) failed but others have been gratifyingly successful. Citrus canker (*Xanthomonas citri*) has been eliminated from the large orchard areas at Kerikeri and Otumoetai, but there are still small pockets of infection in Taranaki, while an outbreak has recently been located in Auckland. Again, though onion yellow dwarf, a serious virus disease of onions, was well established by 1939 in the onion-growing area of Marshland, Christchurch, it has not been observed there for the last 20 years. Regulations enforcing sanitation and prohibiting the growing of onion-seed crops, and alternative hosts brought about the eradication of this disease from the district.

Even if not completely eliminated from the country, onion yellow dwarf has been successfully kept out of the major onion-growing areas. Attempts to eradicate *Verticillium cinerescens* of carnations, and the virus diseases Moorpark mottle of apricots and plum fruit crinkle appear to have been successful. Hot water treatment of barley, introduced during the 1920s, together with the production of mother seed and certification of seed crops, has eliminated the serious diseases of loose smut (*Ustilago tritici*) and covered smut (*Ustilago hordei*) from our malting barleys. There are many ways of keeping diseases in check. Their incidence and severity may be greatly influenced by cultural factors such as soil type, date of sowing, size of plantings, crop rotation, sanitation, etc. Knowledge of these factors is used to reduce losses. Perhaps the most satisfactory way is to use resistant varieties, to the selection and breeding of which much attention has been given, with marked success in increased production of field crops and vegetables.

Heavy losses can occur through the use of diseased stocks of vegetatively propagated plants. A seed potato certification scheme introduced in 1928 has done much to increase potato yields. Similarly, the health of fruit-tree nursery stock has been greatly improved by selection of disease-free root stocks and scion wood. Work begun by New Zealand pathologists in the 1920s contributed much to knowledge of seed-borne diseases and their control. Since that date the general application of seed treatment to many field and vegetable crops has been of considerable benefit to the country's economy. There are some serious soilborne diseases particularly dangerous in glasshouses. Successful investigations on soil treatment have led to control of many of the more important of these, thereby achieving much heavier yields of tomatoes and other crops. Treatments have also been evolved and successfully applied for control of soilborne organisms in the field.

Control of the majority of fungous and bacterial diseases is achieved by the application of therapeutant sprays. A great deal of research work has gone into finding the most effective material for specific diseases, the optimum concentrations, and the times and number of applications needed. As materials of high quality are essential for best results, the extent of control was greatly improved in 1938 by the introduction of a therapeutant certification scheme; but with increasing numbers of therapeutants becoming available it was not possible with limited scientific resources to maintain this. In 1959 it was replaced by an approval and registration scheme administered by an Agricultural Chemicals Board.

Literature on Plant Diseases

New Zealand has a comprehensive literature on plant diseases and their control. Most of the earlier papers appeared in annual reports of the Department of Agriculture and in the *New Zealand Journal of Agriculture*. For the past 25 years, however, the more technical papers have been published in the *New Zealand Journal of Science and Technology* and, more recently, in the *New Zealand Journal of Agricultural Research*. Popular papers have continued to appear in the *New Zealand Journal of Agriculture* and also in the *Orchardist of New Zealand*, the *New Zealand Commercial Grower*, and other growers' journals. Articles on the classification of fungi have been published, mainly in the *Transactions and Proceedings of the Royal Society of New Zealand*, and some papers have appeared in overseas journals.

Chapter 7

Bug Biography: Bacteria that Promote Plant Growth

Certain strains of the soil bacteria Pseudomonas fluorescens have anti-fungal activity that inhibits some plant pathogens. P. fluorescens and other Pseudomonas and Xanthomonas species can increase plant growth in several ways. They may produce a compound that inhibits the growth of pathogens or reduces invasion of the plant by a pathogen.

They may also produce compounds (growth factors) that directly increase plant growth. These plant growth-enhancing bacteria occur naturally in soils, but not always in high enough numbers to have a dramatic effect. In the future, farmers may be able to inoculate seeds with anti-fungal bacteria, such as P. fluorescens, to ensure that the bacteria reduce pathogens around the seed and root of the crop. It's easy to think of all microorganisms as bad things, as germs. But many microorganisms are very helpful, and some are even essential. For example, in agriculture, microorganisms are used to fight pest insects, diseases and weeds that make producing crops and raising livestock less efficient. Other microorganisms help make nutrients in the soil more available to plants.

What Happens to the Way Plants Grow if there are no Microorganisms in the Soil?

Take a sample of fertile soil from a field or garden and divide it into two portions. Bake one in an oven (to destroy the microorganisms). Leave the other portion alone as a control. Plant the same number of seeds in each soil sample. Remember to treat both samples the same while the plants are growing.

Make sure all the plants receive the same amounts of water and light, and are kept at the same temperature. How do the plants differ as they grow?

Are Different Plants Affected in Different Ways by Specific Microorganisms?

Some microorganisms and plants form mutually beneficial partnerships. For example, certain bacteria make a natural nitrogen fertilizer for plants in the family called legumes. Peas, alfalfa and soybeans are legumes. The nitrogen-fixing bacteria are available from garden supply stores and by mail order. Grow both legumes and non-legume plants with and without the bacteria. Are there differences in how well the plants grow?

Suppressive Soil

Suppressive soils hold considerable potential for managing soilborne pathogens. When the suppressiveness has a biological origin, identifying the causal organisms is the crucial step in realising this potential. Armed with such knowledge, it may be possible to develop effective and sustainable pest management strategies through application of these organisms or agronomic practices that influence their population densities.

This chapter focuses on the development and utilisation of a population-based approach for identifying microorganisms involved in specific pathogen suppression. Key experimental design principles of the approach are explored by examining experiments characterising the biological nature of take-all decline. We also describe how this approach was used to identify microorganisms that suppress the sugarbeet cyst nematode. Additional experimental design considerations and future directions for such investigations are also discussed.

Soil Quality Concepts

Soil quality is how well soil does what we want it to do. More specifically, soil quality is the capacity of a specific kind of soil to function, within natural or managed ecosystem boundaries, to sustain plant and animal productivity, maintain or enhance water and air quality, and support human health and habitation. Soil Organic Matter and Soil Biology play a major role in soil quality.

People have different ideas of what a quality soil is. For example:

- for people active in production agriculture, it may mean highly productive land, sustaining or enhancing productivity, maximising profits, or maintaining the soil resource for future generations;
- for consumers, it may mean plentiful, healthful, and inexpensive food for present and future generations;
- for naturalists, it may mean soil in harmony with the landscape and its surroundings;
- for the environmentalist, it may mean soil functioning at its potential in an ecosystem with respect to maintenance or enhancement of biodiversity, water quality, nutrient cycling, and biomass production.

What Soil Does

Healthy soil gives us clean air and water, bountiful crops and forests, productive grazing lands, diverse wildlife, and beautiful landscapes. Soil does all this by performing five essential functions:

- Regulating water - Soil helps control where rain, snowmelt, and irrigation water goes. Water and dissolved solutes flow over the land or into and through the soil.
- Sustaining plant and animal life - The diversity and productivity of living things depends on soil.
- Filtering potential pollutants - The minerals and microbes in soil are responsible for filtering, buffering, degrading, immobilising, and detoxifying organic and inorganic materials, including industrial and municipal by-products and atmospheric deposits.
- Cycling nutrients - Carbon, nitrogen, phosphorus, and many other nutrients are stored, transformed, and cycled in the soil.
- Supporting structures - Buildings need stable soil for support, and archeological treasures associated with human habitation are protected in soils.

The Inherent and Dynamic Qualities of Soil

Soil has both inherent and dynamic qualities. Inherent soil quality is a soil's natural ability to function. For example, sandy soil drains faster than clayey soil. Deep soil has more room for roots than soils with bedrock near the surface. These characteristics do not change easily. Dynamic soil quality is how soil changes depending on how it is managed. Management choices affect the amount of soil organic matter, soil structure, soil depth, and water and nutrient holding capacity. One goal of soil quality research is to learn how to manage soil in a way that improves soil function. Soils respond differently to management depending on the inherent properties of the soil and the surrounding landscape.

Soil Quality Link to Sustainability

Understanding soil quality means assessing and managing soil so that it functions optimally now and is not degraded for future use. By monitoring changes in soil quality, a land manager can determine if a set of practices is sustainable. Go to Soil Quality Assessment and Soil Quality Management for more information.

Research Potential

Most soil quality research is motivated by one of two goals:

1. improving land management on farms and watersheds, or
2. monitoring soil at a national or regional scale.

The first goal involves site-specific assessment and decision-making, so the link between researchers and farmers is important to the success of the research. Most research attempts to identify the links among management practices, observable soil characteristics (i.e., soil quality indicators), soil processes (e.g., nutrient cycling), and the performance of soil functions (e.g. productivity and environmental quality). A single study may examine only one or two of these links. Some important directions for future research include:

- measuring the spatial and temporal variability of soil characteristics, and using patterns of variability as an indicator of soil quality,

- further defining the characteristics of a healthy soil biological community, and approaches to managing soil biology,
- describing and managing changes during the transition time when farmers switch from one set of practices to another,
- improving nutrient cycling by managing soil biology, and
- identifying low-cost remote techniques for monitoring soil quality regionally.

Soil Quality is Not an End in Itself

The ultimate purpose of researching and assessing soil quality is not to achieve high aggregate stability, biological activity, or some other soil property. The purpose is to protect and improve long-term agricultural productivity, water quality, and habitats of all organisms, including people. We use soil characteristics as indicators of soil quality, but in the end, soil quality must be identified by how soil performs its functions.

Soil Biology and the Landscape

An incredible diversity of organisms make up the soil food web. They range in size from the tiniest one-celled bacteria, algae, fungi, and protozoa, to the more complex nematodes and micro-arthropods, to the visible earthworms, insects, small vertebrates, and plants.

As these organisms eat, grow, and move through the soil, they make it possible to have clean water, clean air, healthy plants, and moderated water flow. There are many ways that the soil food web is an integral part of landscape processes. Soil organisms decompose organic compounds, including manure, plant residue, and pesticides, preventing them from entering water and becoming pollutants.

They sequester nitrogen and other nutrients that might otherwise enter groundwater, and they fix nitrogen from the atmosphere, making it available to plants. Many organisms enhance soil aggregation and porosity, thus increasing infiltration and reducing runoff. Soil organisms prey on crop pests and are food for above-ground animals.

The Food Web: Organisms and their Interaction

The soil food web is the community of organisms living all or part of their lives in the soil. A food web diagram shows a series of conversions (represented by arrows) of energy and nutrients as one organism eats another.

All food webs are fuelled by the primary producers: the plants, lichens, moss, photosynthetic bacteria, and algae that use the sun's energy to fix carbon dioxide from the atmosphere. Most other soil organisms get energy and carbon by consuming the organic compounds found in plants, other organisms, and waste by-products. A few bacteria, called chemoautotrophs, get energy from nitrogen, sulfur, or iron compounds rather than carbon compounds or the sun.

As organisms decompose complex materials, or consume other organisms, nutrients are converted from one form to another, and are made available to plants and to other soil organisms. All plants – grass, trees, shrubs, agricultural crops – depend on the food web for their nutrition.

What do Soil Organisms Do?

Growing and reproducing are the primary activities of all living organisms. As individual plants and soil organisms work to survive, they depend on interactions with each other. By-products from growing roots and plant residue feed soil organisms. In turn, soil organisms support plant health as they decompose organic matter, cycle nutrients, enhance soil structure, and control the populations of soil organisms including crop pests.

Organic Matter Fuels the Food Web

Organic matter is many different kinds of compounds – some more useful to organisms than others. In general, soil organic matter is made of roughly equal parts humus and active organic matter. Active organic matter is the portion available to soil organisms. Bacteria tend to use simpler organic compounds, such as root exudates or fresh plant residue. Fungi tend to use more complex compounds, such as fibrous plant residues, wood and soil humus.

Intensive tillage triggers spurts of activity among bacteria and other organisms that consume organic matter (convert it to CO2), depleting the active fraction first. Practices that build soil organic matter (reduced tillage and regular additions of organic material) will raise the proportion of active organic matter long before increases in total organic matter can be measured.

As soil organic matter levels rise, soil organisms play a role in its conversion to humus—a relatively stable form of carbon sequestered in soils for decades or even centuries.

Food Sources For Soil Organisms

"Soil organic matter" includes all the organic substances in or on the soil. Here are terms used to describe different types of organic matter.

- Living organisms: Bacteria, fungi, nematodes, protozoa, earthworms, arthropods, and living roots.
- Dead plant material; organic material; detritus; surface residue: All these terms refer to plant, animal, or other organic substances that have recently been added to the soil and have only begun to show signs of decay. Detritivores are organisms that feed on such material.
- Active fraction organic matter: Organic compounds that can be used as food by microorganisms. The active fraction changes more quickly than total organic matter in response to management changes.
- Labile organic matter: Organic matter that is easily decomposed.
- Root exudates: Soluble sugars, amino acids and other compounds secreted by roots.
- Particulate organic matter (POM) or Light fraction (LF) organic matter: POM and LF have precise size and weight definitions. They are thought to represent the active fraction of organic matter which is more difficult to define. Because POM or LF is larger and lighter than other types of soil organic matter, they can be separated from soil by size (using a sieve) or by weight (using a centrifuge).
- Lignin: A hard-to-degrade compound that is part of the fibres of older plants. Fungi can use the carbon ring structures in lignin as food.

- Recalcitrant organic matter: Organic matter such as humus or lignin-containing material that few soil organisms can decompose.

Humus or humified organic matter: Complex organic compounds that remain after many organisms have used and transformed the original material. Humus is not readily decomposed because it is either physically protected inside of aggregates or chemically too complex to be used by most organisms. Humus is important in binding tiny soil aggregates, and improves water and nutrient holding capacity.

Soil Organisms are Concentrated

Around roots. The rhizosphere is the narrow region of soil directly around roots. It is teeming with bacteria that feed on sloughed-off plant cells and the proteins and sugars released by roots. The protozoa and nematodes that graze on bacteria are also concentrated near roots. Thus, much of the nutrient cycling and disease suppression needed by plants occurs immediately adjacent to roots. In litter. Fungi are common decomposers of plant litter because litter has large amounts of complex, hard-to-decompose carbon. Fungal hyphae (fine filaments) can "pipe" nitrogen from the underlying soil to the litter layer. Bacteria cannot transport nitrogen over distances, giving fungi an advantage in litter decomposition, particularly when litter is not mixed into the soil profile. However, bacteria are abundant in the green litter of younger plants which is higher in nitrogen and simpler carbon compounds than the litter of older plants. Bacteria and fungi are able to access a larger surface area of plant residue after shredder organisms such as earthworms, leaf-eating insects, millipedes, and other arthropods break up the litter into smaller chunks.

On humus. Fungi are common here. Much organic matter in the soil has already been decomposed many times by bacteria and fungi, and/or passed through the guts of earthworms or arthropods. The resulting humic compounds are complex and have little available nitrogen.

Only fungi make some of the enzymes needed to degrade the complex compounds in humus. On the surface of soil aggregates. Biological activity, in particular that of aerobic bacteria and fungi, is greater near the surfaces of soil aggregates than within aggregates. Within large aggregates, processes that do not require oxygen, such as denitrification, can occur. Many aggregates are actually the fecal pellets of earthworms and other invertebrates. In spaces between soil aggregates. Those arthropods and nematodes that cannot burrow through soil move in the pores between soil aggregates. Organisms that are sensitive to desiccation, such as protozoa and many nematodes, live in water-filled pores.

When are they Active?

The activity of soil organisms follows seasonal patterns, as well as daily patterns. In temperate systems, the greatest activity occurs in late spring when temperature and moisture conditions are optimal for growth. However, certain species are most active in the winter, others during dry periods, and still others in flooded conditions. Not all organisms are active at a particular time.

Even during periods of high activity, only a fraction of the organisms are busily eating, respiring, and altering their environment. The remaining portion are barely active or even dormant. Many different organisms are active at different times, and interact with one another, with plants, and with the soil. The combined result is a number of beneficial functions including nutrient cycling, moderated water flow, and pest control.

The Importance of the Soil Food Web

The living component of soil, the food web, is complex and has different compositions in different ecosystems. Management of croplands, rangelands, forestlands, and gardens benefits from and affects the food web.

The next unit of the Soil Biology Primer, "The Food Web & Soil Health," introduces the relationship of soil biology to agricultural productivity, biodiversity, carbon sequestration and to air and water quality. The remaining six units of the Soil Biology Primer describe the major groups of soil organisms: bacteria, fungi, protozoa, nematodes, arthropods, and earthworms.

For more information about the diversity within each organism group, see the list of readings at the end of "The Food Web & Soil Health" unit.

The Food Web & Soil Health

How Do Food Webs Differ?

Each field, forest, or pasture has a unique soil food web with a particular proportion of bacteria, fungi, and other groups, and a particular level of complexity within each group of organisms. These differences are the result of soil, vegetation, and climate factors, as well as land management practices.

What is Complexity?

Food web complexity is a factor of both the number of species and the number of different kinds of species in the soil. For example, a soil with ten species of bacterial-feeding nematodes is less complex than a soil with ten nematode species that includes bacterial-feeders, fungal-feeders, and predatory nematodes.

Complexity can be determined, in part, from a food, which represents the soil in an old-growth Douglas fir forest. Each box of the food web diagram represents a functional group of organisms that perform similar roles in the soil system. Transfers of energy are represented by the arrows on the diagram and occur when one organism eats another. Complex ecosystems have more functional groups and more energy transfers than simple ecosystems.

The number of functional groups that turn over energy before the energy leaves the soil system is different (and characteristic) for each ecosystem. In the Douglas fir system, energy may undergo more than twenty transfers from organism to organism, or between functional groups. In contrast, a cave or low-residue cultivated system is not likely to include a large variety of higher predators on the right-hand side of a soil food web diagram.

Energy and nutrients will be cycled through fewer types of organisms. Land management practices can alter the number of functional groups – or complexity – in the soil. Intensively managed systems, such as cropland, have varied numbers of functional groups. Crop selections, tillage practices, residue management, pesticide use, and irrigation alter the habitat for soil organisms, and thus alter the structure and complexity of the food web.

Benefits of Complexity

Biological complexity of a soil system can affect processes such as nutrient cycling, the formation of soil structure, pest cycles, and decomposition rates. Researchers have yet to define how much and what kind of food web complexity in managed ecosystems is optimal for these soil processes.

Nutrient cycling. When organisms consume food, they create more of their own biomass and they release wastes. The most important waste for crop growth is ammonium (NH4+). Ammonium and other readily utilised nutrients are quickly taken up by other organisms, including plant roots. When a large variety of organisms are present, nutrients may cycle more rapidly and frequently among forms that plants can and cannot use. Nutrient retention. In addition to mineralising or releasing nitrogen to plants, the soil food web can immobilise or retain nitrogen when plants are not rapidly growing. Nitrogen in the form of soil organic matter and organism biomass is less mobile and less likely to be lost from the rooting zone than inorganic nitrate (NO_3^-) and ammonium (NH_4^+).

Improved structure, infiltration, and water-holding capacity. Many soil organisms are involved in the formation and stability of soil aggregates. Bacterial activity, organic matter, and the chemical properties of clay particles are responsible for creating microaggregates from individual soil particles. Earthworms and arthropods consume small aggregates of mineral particles and organic matter, and generate larger fecal pellets coated with compounds from the gut. These fecal pellets become part of the soil structure. Fungal hyphae and root hairs bind together and help stabilise larger aggregates. Improved aggregate stability, along with the burrows of earthworms and arthropods, increases porosity, water infiltration, and water-holding capacity. Disease suppression. A complex soil food web contains numerous organisms that can compete with disease-causing organisms. These competitors may prevent soil pathogens from establishing on plant surfaces, prevent pathogens from getting food, feed on pathogens, or generate metabolites that are toxic to or inhibit pathogens.

Degradation of pollutants. An important role of soil is to purify water. A complex food web includes organisms that consume (degrade) a wide range of pollutants under a wide range of environmental conditions. Biodiversity. Greater food web complexity means greater biodiversity. Biodiversity is measured by the total number of species, as well as the relative abundance of these species, and the number of functional groups of organisms.

Management and Soil Health

A healthy soil effectively supports plant growth, protects air and water quality, and ensures human and animal health. The physical structure, chemical make-up, and biological components of the soil together determine how well a soil performs these services. In every healthy system or watershed, the soil food web is critical to major soil functions including:

- sustaining biological activity, diversity, and productivity;
- regulating the flow of water and dissolved nutrients;
- storing and cycling nutrients and other elements; and
- filtering, buffering, degrading, immobilising and detoxifying organic and inorganic materials that are potential pollutants.

The interactions among organisms enhance many of these functions.

Successful land management requires approaches that protect all resources, including soil, water, air, plants, animals and humans. Many management strategies change soil habitats and the food web, and alter soil quality, or the capacity of soil to perform its functions. Examples of some practices that change the complexity and health of the soil community include:

- Compared to a field with a 2-year crop rotation, a field with a 4 crops grown in rotation may have a greater variety of food sources (i.e., roots and surface residue), and therefore is likely to have more types of bacteria, fungi, and other organisms.
- A cleanly-tilled field with few vegetated edges may have fewer habitats for arthropods than a field broken up by grassed waterways, terraces, or fence rows.
- Although the effect of pesticides on soil organisms varies, high levels of pesticide use will generally reduce food

web complexity. An extreme example is the repeated use of methyl bromide which has been observed to eliminate most soil organisms except a few bacteria species.

The Food Web and Carbon Sequestration

Land management practices can be chosen to increase the amount of carbon sequestered as soil organic matter and reduce the amount of CO_2, a greenhouse gas, released to the atmosphere. As the soil food web decomposes organic material, it releases carbon into the atmosphere as CO_2 or converts it to a variety of forms of soil organic matter. Labile or active fractions of organic matter stay in the soil for a few years. Stable forms reside in the soil for decades or hundreds of years. Physically stabilised organic matter is protected inside soil aggregates that soil organisms help create. Humified organic matter is stable because bacteria and fungi have helped form molecules that are too complex and large for soil organisms to decompose.

Looking Forward

The functions of the food web are essential to plant growth and environmental quality. Good resource management will integrate food web-enhancing strategies into the regular activities of farms, ranches, forests, or in backyard gardens. Needed research will examine food web functions within whole systems, and will support technology development. Technology to assess and maintain the functions of soil food webs will be developed to assist land managers and researchers as they strive towards soil productivity and stewardship. In the coming years, we can expect progress at answering soil biology questions such as the following.

What is a healthy food web? What measurements or observations can be used to determine whether a particular biological community is desirable for the intended land use? What level of complexity is optimal for highly productive and sustainable crop, range or forest lands?

Is it more useful to count species, or types of organisms? The Soil Biology Primer divides food web organisms into six groups. Achieving an optimal balance of these groups is one approach to managing the food web. Alternatively, identifying the species and complexity present within a group may provide other useful information about the health and productive potential of a soil.

How should the biology of the soil be managed? In the future, land managers may be able to more precisely predict the effect of management decisions such as the timing of tillage, the application of a certain kind of compost, or the use of a particular pesticide. They may choose practices with the intent of making specific changes to the composition of the soil food web.

What are the costs and benefits of managing for soil biological functions? The costs to achieve a highly diverse, or complex, soil community need to be identified. These can be compared to the benefits of biological services provided, such as nutrient cycling, disease suppression, and soil structure enhancement.

Earthworms

Of all the members of the soil food web, earthworms need the least introduction. Most people become familiar with these soft, slimy, invertebrates at a young age. Earthworms are hermaphrodites, meaning that they exhibit both male and female characteristics. They are major decomposers of dead and decomposing organic matter, and derive their nutrition from the bacteria and fungi that grow upon these materials. They fragment organic matter and make major contributions to recycling the nutrients it contains. Earthworms occur in most temperate soils and many tropical soils. They are divided into 23 families, more than 700 genera, and more than 7,000 species. They range from an inch to two yards in length and are found seasonally at all depths in the soil. In terms of biomass and overall activity, earthworms dominate the world of soil invertebrates, including arthropods.

What Do Earthworms Do?

Earthworms dramatically alter soil structure, water movement, nutrient dynamics, and plant growth. They are not essential to all healthy soil systems, but their presence is usually an indicator of a healthy system. Earthworms perform several beneficial functions. Stimulate microbial activity.

Although earthworms derive their nutrition from microorganisms, many more microorganisms are present in their feces or casts than in the organic matter that they consume. As organic matter passes through their intestines, it is fragmented and inoculated with microorganisms. Increased microbial activity facilitates the cycling of nutrients from organic matter and their conversion into forms readily taken up by plants.

Mix and aggregate soil. As they consume organic matter and mineral particles, earthworms excrete wastes in the form of casts, a type of soil aggregate. Charles Darwin calculated that earthworms can move large amounts of soil from the lower strata to the surface and also carry organic matter down into deeper soil layers. A large proportion of soil passes through the guts of earthworms, and they can turn over the top six inches (15 cm) of soil in ten to twenty years. Increase infiltration. Earthworms enhance porosity as they move through the soil. Some species make permanent burrows deep into the soil. These burrows can persist long after the inhabitant has died, and can be a major conduit for soil drainage, particularly under heavy rainfall. At the same time, the burrows minimise surface water erosion. The horizontal burrowing of other species in the top several inches of soil increases overall porosity and drainage.

Improve water-holding capacity. By fragmenting organic matter, and increasing soil porosity and aggregation, earthworms can significantly increase the water-holding capacity of soils. Provide channels for root growth. The channels made by deep-burrowing earthworms are lined with readily available nutrients and make it easier for roots to penetrate deep into the soil. Bury and shred plant residue. Plant and crop residue are gradually buried by cast material deposited on the surface and as earthworms pull surface residue into their burrows.

Where Are Earthworms?

Different species of earthworms inhabit different parts of the soil and have distinct feeding strategies. They can be separated into three major ecological groups based on their feeding and burrowing habits. All three groups are common and important to soil structure. Surface soil and litter species

– Epigeic species. These species live in or near surface plant litter. They are typically small and are adapted to the highly variable moisture and temperature conditions at the soil surface. The worms found in compost piles are epigeic and are unlikely to survive in the low organic matter environment of soil.

Upper soil species – Endogeic species. Some species move and live in the upper soil strata and feed primarily on soil and associated organic matter (geophages). They do not have permanent burrows, and their temporary channels become filled with cast material as they move through the soil, progressively passing it through their intestines.

Deep-burrowing species – Anecic species. These earthworms, which are typified by the "night crawler," *Lumbricus terrestris*, inhabit more or less permanent burrow systems that may extend several metres into the soil. They feed mainly on surface litter that they pull into their burrows. They may leave plugs, organic matter, or cast (excreted soil and mineral particles) blocking the mouth of their burrows.

Looking For Earthworms?

It is easy to determine whether you have an adequate population of earthworms in your soil. Look for their casts in the forms of little piles of soil, mineral particles, or organic matter at the soil surface. They can be seen moving over the soil surface or even breeding, particularly on warm, damp nights. Dump a spade full of moist soil into a bucket or onto a sheet of plastic, and sort through for earthworms. Can you identify different species? To find the deep burrowing species, pour a dilute mustard solution onto the soil. Many will quickly come to the soil surface in response to this irritant.

Abundance and Distribution of Earthworms

The majority of temperate and many tropical soils support significant earthworm populations. A square yard of cropland in the United States can contain from 50-300 earthworms, or even larger populations in highly organic soils. A similar area of grassland or temperate woodlands will have from 100-500 earthworms. Based on their total biomass, earthworms are the predominant group of soil invertebrates in most soils. The family of earthworms that is most important in enhancing agricultural soil is Lumbricidae, which includes the genuses *Lumbricus, Aporrectodea,* and several others. Lumbricids originated in Europe and have been transported by human activities to many parts of the world. The United States has only one or two known native species of lumbricids. Others were brought to this country by settlers (probably in potted plants from Europe), and were distributed down the waterways.

Generally, lumbricids are much more common in the north and east than in the drier south and west of the United States. They tend to be more abundant in loam and clay loam and even in silty soil, than in sandy soil and heavy clay. Populations also build up in irrigated soil. Earthworm populations tend to increase with soil organic matter levels and decrease with soil disturbances, such as tillage and potentially harmful chemicals.

Interactions of Earthworms with other Members of the Food Web

The lives of earthworms and microbes are closely intertwined. Earthworms derive their nutrition from fungi, bacteria, and possibly protozoa and nematodes, and they promote the activity of these organisms by shredding and increasing the surface area of organic matter and making it more available to small organisms.

Earthworms also influence other soil-inhabiting invertebrates by changing the amount and distribution of organic matter and microbial populations. There is good evidence that earthworm activity affects the spatial distribution of soil microarthropod communities in the soil. Earthworms have few invertebrate enemies, other than flatworms and a species of parasitic fly. Their main predators are a wide range of birds and mammals that prey upon them at the soil surface.

Earthworms and Water Quality

Earthworms improve water infiltration and water holding capacity because their shredding, mixing, and defecating enhances soil structure. In addition, burrows provide quick entry for water into and through soil. High infiltration rates help prevent pollution by minimising runoff, erosion, and chemical transport to surface waters. There is concern that burrows may increase the transport of pollutants, such as nitrates or pesticides, into groundwater. However, the movement of potential pollutants through soil is not a straightforward process and it is not clear when earthworm activity will or will not have a negative impact on groundwater quality.

Whether pollutants reach groundwater depends on a number of factors, including the location of pollutants on the surface or within soil, the quantity and intensity of rain, how well water moves into and through other parts of the soil, and characteristics of the burrows. The horizontal burrows of endogeic earthworms (such as *Aporrectodea tuberculata*, which are common in Midwestern fields) do not transport water and solutes as deeply as the vertical burrows of night crawlers (*L. terrestris*) and other anecic species. Even vertical burrows, however, are not direct channels for water movement. They have bends and turns and are lined with organic matter that adsorbs many potential pollutants from the water. Although there is much more to learn about how earthworms affect water movement through soil, they clearly help minimise pollution of surface waters by improving infiltration rates and decreasing runoff.

Bug Biography: Night Crawlers and Tillage

The substitution of conventional tillage by no-till or conservation tillage is increasingly common and widely adopted in the United States and elsewhere. In these situations, earthworms, particularly the "night crawler," *Lumbricus terrestris* L., are especially important. Earthworms become the main agent for incorporating crop residue into the soil by pulling some into their burrows and by slowly burying the remainder under casts laid on the soil surface. In reduced tillage systems, surface residue builds up and triggers growth in earthworm populations. Earthworms need the food and habitat provided by surface residue, and they eat the fungi that become more common in no-till soils. As earthworm populations increase, they pull more and more residue into their burrows, helping to mix organic matter into the soil, improving soil structure and water infiltration.

Arthropods

Many bugs, known as arthropods, make their home in the soil. They get their name from their jointed (arthros) legs (podos). Arthropods are invertebrates, that is, they have no backbone, and rely instead on an external covering called an exoskeleton. Arthropods range in size from microscopic to several inches in length. They include insects, such as springtails, beetles, and ants; crustaceans such as sowbugs; arachnids such as spiders and mites; myriapods, such as centipedes and millipedes; and scorpions.

Nearly every soil is home to many different arthropod species. Certain row-crop soils contain several dozen species of arthropods in a square mile. Several thousand different species may live in a square mile of forest soil. Arthropods can be grouped as shredders, predators, herbivores, and fungal-feeders, based on their functions in soil. Most soil-dwelling arthropods eat fungi, worms, or other arthropods. Root-feeders and dead-plant shredders are less abundant. As they feed, arthropods aerate and mix the soil, regulate the population size of other soil organisms, and shred organic material.

Predators

Predators and micropredators can be either generalists, feeding on many different prey types, or specialists, hunting only a single prey type. Predators include centipedes, spiders, ground-beetles, scorpions, skunk-spiders, pseudoscorpions, ants, and some mites. Many predators eat crop pests, and some, such as beetles and parasitic wasps, have been developed for use as commercial biocontrols.

Herbivores

Numerous root-feeding insects, such as cicadas, mole-crickets, and anthomyiid flies (root-maggots), live part of all of their life in the soil. Some herbivores, including rootworms and symphylans, can be crop pests where they occur in large numbers, feeding on roots or other plant parts.

Fungal Feeders

Arthropods that graze on fungi (and to some extent bacteria) include most springtails, some mites, and silverfish. They scrape and consume bacteria and fungi off root surfaces. A large fraction of the nutrients available to plants is a result of microbial-grazing and nutrient release by fauna.

What is in Your Soil?

If you would like to see what kind of organisms are in your soil, you can easily make a pitfall trap to catch large arthropods, and a Burlese funnel to catch small arthropods. Make a pitfall trap by sinking a pint- or quart-sized container (such as a yogurt cup) into the ground so the rim is level with the soil surface. If desired, fashion a roof over the cup to keep the rain out, and add 1/2 of an inch of non-hazardous antifreeze to the cup to preserve the creatures and prevent them from eating one another. Leave in place for a week and wait for soil organisms to fall into the trap.

To make a Burlese funnel, set a piece of 1/4 inch rigid wire screen in the bottom of a funnel to support the soil. (A funnel can be made by cutting the bottom off a plastic soda bottle.) Half fill the funnel with soil, and suspend it over a cup with a bit of anti-freeze or ethyl alcohol in the bottom as a preservative. Suspend a light bulb about 4 inches over the soil to drive the organisms out of the soil and into the cup. Leave the light bulb on for about 3 days to dry out the soil. Then pour the alcohol into a shallow dish and use a magnifying glass to examine the organisms.

What Do Arthropods Do?

Although the plant feeders can become pests, most arthropods perform beneficial functions in the soil-plant system. Shred organic material. Arthropods increase the surface area accessible to microbial attack by shredding dead plant residue and burrowing into coarse woody debris. Without shredders, a bacterium in leaf litter would be like a person in a pantry without a can-opener – eating would be a very slow process. The shredders act like can-openers and greatly increase the rate of decomposition. Arthropods ingest decaying plant material to eat the bacteria and fungi on the surface of the organic material. Stimulate microbial activity. As arthropods graze on bacteria and fungi, they stimulate the growth of mycorrhizae and other fungi, and the decomposition of organic matter. If grazer populations get too dense the opposite effect can occur – populations of bacteria and fungi will decline. Predatory arthropods are important to keep grazer populations under control and to prevent them from over-grazing microbes.

Mix microbes with their food. From a bacterium's point-of-view, just a fraction of a millimetre is infinitely far away. Bacteria have limited mobility in soil and a competitor is likely to be closer to a nutrient treasure. Arthropods help out by distributing nutrients through the soil, and by carrying bacteria on their exoskeleton and through their digestive system. By more thoroughly mixing microbes with their food, arthropods enhance organic matter decomposition. Mineralise plant nutrients. As they graze, arthropods mineralise some of the nutrients in bacteria and fungi, and excrete nutrients in plant-available forms.

Enhance soil aggregation. In most forested and grassland soils, every particle in the upper several inches of soil has been through the gut of numerous soil fauna. Each time soil passes through another arthropod or earthworm, it is thoroughly mixed with organic matter and mucus and deposited as fecal pellets. Fecal pellets are a highly concentrated nutrient resource, and are a mixture of the organic and inorganic substances required for growth of bacteria and fungi. In many soils, aggregates between 1/10,000 and 1/10 of an inch (0.0025mm and 2.5mm) are actually fecal pellets.

Burrow. Relatively few arthropod species burrow through the soil. Yet, within any soil community, burrowing arthropods and earthworms exert an enormous influence on the composition of the total fauna by shaping habitat. Burrowing changes the physical properties of soil, including porosity, water-infiltration rate, and bulk density. Stimulate the succession of species. A dizzying array of natural bio-organic chemicals permeates the soil. Complete digestion of these chemicals requires a series of many types of bacteria, fungi, and other organisms with different enzymes. At any time, only a small subset of species is metabolically active – only those capable of using the resources currently available. Soil arthropods consume the dominant organisms and permit other species to move in and take their place, thus facilitating the progressive breakdown of soil organic matter.

Control pests. Some arthropods can be damaging to crop yields, but many others that are present in all soils eat or compete with various root- and foliage-feeders. Some (the specialists) feed on only a single type of prey species. Other arthropods (the generalists), such as many species of centipedes, spiders, ground-beetles, rove-beetles, and gamasid mites, feed on a broad range of prey. Where a healthy population of generalist predators is present, they will be available to deal with a variety of pest outbreaks.

A population of predators can only be maintained between pest outbreaks if there is a constant source of non-pest prey to eat. That is, there must be a healthy and diverse food web. A fundamental dilemma in pest control is that tillage and insecticide application have enormous effects on non- target species in the food web. Intense land use (especially monoculture, tillage, and pesticides) depletes soil diversity. As total soil diversity declines, predator populations drop sharply and the possibility for subsequent pest outbreaks increases.

Where Do Arthropods Live?

The abundance and diversity of soil fauna diminishes significantly with soil depth. The great majority of all soil species are confined to the top three inches. Most of these creatures have limited mobility, and are probably capable of "cryptobiosis," a state of "suspended animation" that helps them survive extremes of temperature, wetness, or dryness that would otherwise be lethal.

As a general rule, larger species are active on the soil surface, seeking temporary refuge under vegetation, plant residue, wood, or rocks. Many of these arthropods commute daily to forage within herbaceous vegetation above, or even high in the canopy of trees. (For instance, one of these tree-climbers is the caterpillar-searcher used by foresters to control gypsy moth).

Some large species capable of true burrowing live within the deeper layers of the soil. Below about two inches in the soil, fauna are generally small – 1/250 to 1/10 of an inch. (Twenty-five of the smallest of these would fit in a period on this page.) These species are usually blind and lack prominent colouration. They are capable of squeezing through minute pore spaces and along root channels. Subsurface soil dwellers are associated primarily with the rhizosphere (the soil volume immediately adjacent to roots).

Abundance of Arthropods

A single square yard of soil will contain 500 to 200,000 individual arthropods, depending upon the soil type, plant community, and management system. Despite these large numbers, the biomass of arthropods in soil is far less than that of protozoa and nematodes. In most environments, the most abundant soil dwellers are springtails and mites, though ants and termites predominate in certain situations, especially in desert and tropical soils. The largest number of arthropods are in natural plant communities with few earthworms (such as conifer forests). Natural communities with numerous earthworms (such as grassland soils) have the fewest arthropods. Apparently, earthworms out-compete arthropods, perhaps by excessively reworking their habitat or eating them incidentally. However, within pastures and farm lands arthropod numbers and diversity are generally thought to increase as earthworm populations rise. Burrowing earthworms probably create habitat space for arthropods in agricultural soils.

Bug Biography: Springtails

Springtails are the most abundant arthropods in many agricultural and rangeland soils. Populations of tens of thousands per square yard are frequent. When foraging, springtails walk with 3 pairs of legs like most insects, and hold their tail tightly tucked under the belly. If attacked by a predator, body fluid rushes into the tail base, forcing the tail to slam down and catapult the springtail as much as a yard away. Springtails have been shown to be beneficial to crop plants by releasing nutrients and by feeding upon diseases caused by fungi.

Nematodes

Nematodes are non-segmented worms typically 1/500 of an inch (50 μm) in diametre and 1/20 of an inch (1 mm) in length. Those few species responsible for plant diseases have received a lot of attention, but far less is known about the majority of the nematode community that plays beneficial roles in soil. An incredible variety of nematodes function at several trophic levels of the soil food web. Some feed on the plants and algae (first trophic level); others are grazers that feed on bacteria and fungi (second trophic level); and some feed on other nematodes (higher trophic levels).

Free-living nematodes can be divided into four broad groups based on their diet.

- *Bacterial-feeders* consume bacteria.
- *Fungal-feeders* feed by puncturing the cell wall of fungi and sucking out the internal contents.
- *Predatory nematodes* eat all types of nematodes and protozoa. They eat smaller organisms whole, or attach themselves to the cuticle of larger nematodes, scraping away until the prey's internal body parts can be extracted.
- *Omnivores* eat a variety of organisms or may have a different diet at each life stage. *Root-feeders* are plant parasites, and thus are not free-living in the soil.

What Do Nematodes Do?

Nutrient cycling. Like protozoa, nematodes are important in mineralising, or releasing, nutrients in plant-available forms. When nematodes eat bacteria or fungi, ammonium (NH4+) is released because bacteria and fungi contain much more nitrogen than the nematodes require. Grazing. At low nematode densities, feeding by nematodes stimulates the growth rate of prey populations. That is, bacterial-feeders stimulate bacterial growth, plant-feeders stimulate plant growth, and so on. At higher densities, nematodes will reduce the population of their prey. This may decrease plant productivity, may negatively impact mycorrhizal fungi, and can reduce decomposition and immobilisation rates by bacteria and fungi. Predatory nematodes may regulate populations of bacterial-and fungal-feeding nematodes, thus preventing over-grazing by those groups. Nematode grazing may control the balance between bacteria and fungi, and the species composition of the microbial community.

Dispersal of microbes. Nematodes help distribute bacteria and fungi through the soil and along roots by carrying live and dormant microbes on their surfaces and in their digestive systems. Food source. Nematodes are food for higher level predators, including predatory nematodes, soil microarthropods, and soil insects. They are also parasitised by bacteria and fungi. Disease suppression and development. Some nematodes cause disease. Others consume disease-causing organisms, such as root-feeding nematodes, or prevent their access to roots. These may be potential biocontrol agents.

Where are Nematodes?

Nematodes are concentrated near their prey groups. Bacterial-feeders abound near roots where bacteria congregate; fungal-feeders are near fungal biomass; root-feeders are concentrated around roots of stressed or susceptible plants. Predatory nematodes are more likely to be abundant in soils with high numbers of nematodes.

Because of their size, nematodes tend to be more common in coarser-textured soils. Nematodes move in water films in large (>1/500 inch or 50 μm) pore spaces. Agricultural soils generally support less than 100 nematodes in each teaspoon (dry gram) of soil. Grasslands may contain 50 to 500 nematodes, and forest soils generally hold several hundred per teaspoon. The proportion of bacterial-feeding and fungal-feeding nematodes is related to the amount of bacteria and fungi in the soil. Commonly, less disturbed soils contain more predatory nematodes, suggesting that predatory nematodes are highly sensitive to a wide range of disturbances.

Nematodes and Soil Quality

Nematodes may be useful indicators of soil quality because of their tremendous diversity and their participation in many functions at different levels of the soil food web. Several researchers have proposed approaches to assessing the status of soil quality by counting the number of nematodes in different families or trophic groups.* In addition to their diversity, nematodes may be useful indicators because their populations are relatively stable in response to changes in moisture and temperature (in contrast to bacteria), yet nematode populations respond to land management changes in predictable ways. Because they are quite small and live in water films, changes in nematode populations reflect changes in soil microenvironments.

Bug Biography: Nematode Trappers

One group of fungi may be a useful biological control agent against parasitic nematodes. These predatory fungi grow through the soil, setting out traps when they detect signs of their prey. Some species use sticky traps, others make circular rings of hyphae to constrict their prey. When the trap is set, the fungi put out a lure, attracting nematodes that are looking for lunch. The nematode, however, becomes lunch for the fungus.

Protozoa

Protozoa are single-celled animals that feed primarily on bacteria, but also eat other protozoa, soluble organic matter, and sometimes fungi. They are several times larger than bacteria – ranging from 1/5000 to 1/50 of an inch (5 to 500 μm) in diametre. As they eat bacteria, protozoa release excess nitrogen that can then be used by plants and other members of the food web. Protozoa are classified into three groups based on their shape: *Ciliates* are the largest and move by means of hair-like cilia. They eat the other two types of protozoa, as well as bacteria. *Amoebae* also can be quite large and move by means of a temporary foot or "pseudopod." Amoebae are further divided into *testate amoebae* (which make a shell-like covering) and *naked amoebae* (without a covering). *Flagellates* are the smallest of the protozoa and use a few whip-like flagella to move.

What Do Protozoa Do?

Protozoa play an important role in mineralising nutrients, making them available for use by plants and other soil organisms. Protozoa (and nematodes) have a lower concentration of nitrogen in their cells than the bacteria they eat. (The ratio of carbon to nitrogen for protozoa is 10:1 or much more and 3:1 to 10:1 for bacteria.) Bacteria eaten by protozoa contain too much nitrogen for the amount of carbon protozoa need. They release the excess nitrogen in the form of ammonium (NH4+). This usually occurs near the root system of a plant. Bacteria and other organisms rapidly take up most of the ammonium, but some is used by the plant.

Another role that protozoa play is in regulating bacteria populations. When they graze on bacteria, protozoa stimulate growth of the bacterial population (and, in turn, decomposition rates and soil aggregation.) Exactly why this happens is under some debate, but grazing can be thought of like pruning a tree – a small amount enhances growth, too much reduces growth or will modify the mix of species in the bacterial community. Protozoa are also an important food source for other soil organisms and help to suppress disease by competing with or feeding on pathogens.

Where are Protozoa?

Protozoa need bacteria to eat and water in which to move, so moisture plays a big role in determining which types of protozoa will be present and active. Like bacteria, protozoa are particularly active in the rhizosphere next to roots. Typical numbers of protozoa in soil vary widely – from a thousand per teaspoon in low fertility soils to a million per teaspoon in some highly fertile soils. Fungal-dominated soils (e.g. forests) tend to have more testate amoebae and ciliates than other types. In bacterial-dominated soils, flagellates and naked amoebae predominate. In general, high clay-content soils contain a higher number of smaller protozoa (flagellates and naked amoebae), while coarser textured soils contain more large flagellates, amoebae of both varieties, and ciliates.

Nematodes and Protozoa

Protozoa and bacterial-feeding nematodes compete for their common food resource: bacteria. Some soils have high numbers of either nematodes or protozoa, but not both. The significance of this difference to plants is not known. Both groups consume bacteria and release NH_4^+.

Bug Biography: Soil Dwelling Vampires

Most protozoa eat bacteria, but one group of amoebae, the vampyrellids, eat fungi. The perfectly round holes drilled through the fungal cell wall, much like the purported puncture marks on the neck of a vampire's victim, are evidence of the presence of vampyrellid amoebae. The amoebae attach to the surface of fungal hyphae and generate enzymes that eat through the fungal cell wall. The amoeba then sucks dry or engulfs the cytoplasm inside the fungal cell before moving on to its next victim. Vampyrellids attack many fungi including root pathogens, such as Gaeumannomyces graminis, shown in the photo. This fungus attacks wheat roots and causes take-all disease.

Fungi

Fungi are microscopic cells that usually grow as long threads or strands called hyphae, which push their way between soil particles, roots, and rocks. Hyphae are usually only several thousandths of an inch (a few micrometres) in diametre. A single hyphae can span in length from a few cells to many yards. A few fungi, such as yeast, are single cells. Hyphae sometimes group into masses called mycelium or thick, cord-like "rhizomorphs" that look like roots. Fungal fruiting structures (mushrooms) are made of hyphal strands, spores, and some special structures like gills on which spores form. A single individual fungus can include many fruiting bodies scattered across an area as large as a baseball diamond.

Fungi perform important services related to water dynamics, nutrient cycling, and disease suppression. Along with bacteria, fungi are important as decomposers in the soil food web. They convert hard-to-digest organic material into forms that other organisms can use. Fungal hyphae physically bind soil particles together, creating stable aggregates that help increase water infiltration and soil water holding capacity. Soil fungi can be grouped into three general functional groups based on how they get their energy.

- *Decomposers* – saprophytic fungi – convert dead organic material into fungal biomass, carbon dioxide (CO2), and small molecules, such as organic acids. These fungi generally use complex substrates, such as the cellulose and lignin, in wood, and are essential in decomposing the carbon ring structures in some pollutants. A few fungi are called "sugar fungi" because they use the same simple substrates as do many bacteria. Like bacteria, fungi are important for immobilising, or retaining, nutrients in the soil. In addition, many of the secondary metabolites of fungi are organic acids, so they help increase the accumulation of humic-acid rich organic matter that is resistant to degradation and may stay in the soil for hundreds of years.
- *Mutualists* – the mycorrhizal fungi – colonise plant roots. In exchange for carbon from the plant, mycorrhizal fungi help solubolize phosphorus and bring soil nutrients

(phosphorus, nitrogen, micronutrients, and perhaps water) to the plant. One major group of mycorrhizae, the *ectomycorrhizae*, grow on the surface layers of the roots and are commonly associated with trees. The second major group of mycorrhizae are the *endomycorrhizae* that grow within the root cells and are commonly associated with grasses, row crops, vegetables, and shrubs. Arbuscular mycorrhizal (AM) fungi are a type of endomycorrhizal fungi. Ericoid mycorrhizal fungi can by either ecto- or endomycorrhizal.

- The third group of fungi, *pathogens* or *parasites*, cause reduced production or death when they colonise roots and other organisms. Root-pathogenic fungi, such as *Verticillium*, *Pythium*, and *Rhizoctonia*, cause major economic losses in agriculture each year. Many fungi help control diseases. For example, nematode-trapping fungi that parasitise disease-causing nematodes, and fungi that feed on insects may be useful as biocontrol agents.

Mycorrhizal Fungi in Agriculture

Mycorrhiza is a symbiotic association between fungi and plant roots and is unlike either fungi or roots alone. Most trees and agricultural crops depend on or benefit substantially from mycorrhizae. The exceptions are many members of the Cruciferae family (e.g., broccoli, mustard), and the Chenopodiaceae family (e.g. lambsquarters, spinach, beets), which do not form mycorrhizal associations.

The level of dependency on mycorrhizae varies greatly among varieties of some crops, including wheat and corn. Land management practices affect the formation of mycorrhizae.

The number of mycorrhizal fungi in soil will decline in fallowed fields or in those planted to crops that do not form mycorrhizae.

Frequent tillage may reduce mycorrhizal associations, and broad spectrum fungicides are toxic to mycorrhizal fungi. Very high levels of nitrogen or phosphorus fertilizer may reduce inoculation of roots. Some inoculums of mycorrhizal fungi are commercially available and can be added to the soil at planting time.

Practices that Influence the Amount of Organic Matter

Human Interventions that Influence Soil Organic Matter

Various types of human activity decrease soil organic matter contents and biological activity. However, increasing the organic matter content of soils or even maintaining good levels requires a sustained effort that includes returning organic materials to soils and rotations with high-residue crops and deep- or dense-rooting crops. It is especially difficult to raise the organic matter content of soils that are well aerated, such as coarse sands, and soils in warm-hot and arid regions because the added materials decompose rapidly. Soil organic matter levels can be maintained with less organic residue in finetextured soils in cold temperate and moist-wet regions with restricted aeration.

Practices that Decrease Soil Organic Matter

Any form of human intervention influences the activity of soil organisms (Curry and Good, 1992) and thus the equilibrium of the system.

Figure: *Severe soil erosion removes the potential energy source for soil microbes, resulting in the death of the* microbial population and thus of the soil itself.

Management practices that alter the living and nutrient conditions of soil organisms, such as repetitive tillage or burning of vegetation, result in a degradation of their microenvironments.

In turn, this results in a reduction of soil biota, both in biomass and diversity. Where there are no longer organisms to decompose soil organic matter and bind soil particles, the soil structure is damaged easily by rain, wind and sun. This can lead to rainwater runoff and soil erosion (Plate 3), removing the potential food for organisms, i.e. the organic matter of the topsoil. Therefore, soil biota are the most important property of the soil, and "when devoid of its biota, the uppermost layer of earth ceases to be soil" (Lal, 1991).

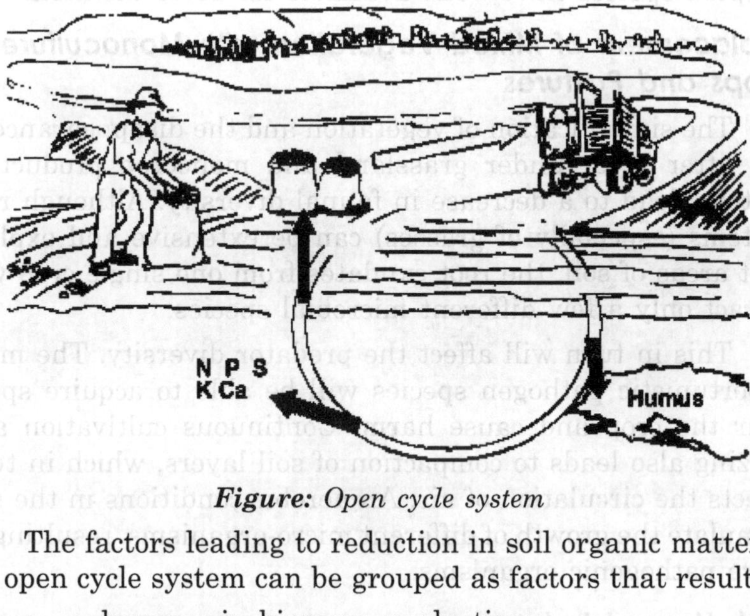

Figure: Open cycle system

The factors leading to reduction in soil organic matter in an open cycle system can be grouped as factors that result in:

- a decrease in biomass production;
- a decrease in organic matter supply;
- increased decomposition rates.

Decrease in Biomass Production

Replacement of Perennial Vegetation

A consequence of clearing forest for agriculture is the disappearance of the litter layer, with a consequent reduction in the numbers and variety of soil organisms. While many temperate forest species appear to adapt well to grassland (Curry and Good, 1992), the effects of deforestation in the tropics appear to be more marked (Plate 4). Studies have shown that as soil biodiversity declines, adapted species may take over from the indigenous species and the composition may change drastically. Soil macrofaunal biomass and population density fell to 6 and 17 percent, respectively, in cultivated plots, compared with primary forest in Peruvian Amazonia (Lavelle and Pashanasi, 1989).

In Suriname, the number of animals per square metre has fallen to 36 percent and the diversity of species has fallen to 28 percent compared with primary forest (Van der Werff, 1990). The indigenous species have largely disappeared, but adapted species have been available for recolonisation.

Replacement of Mixed Vegetation with Monoculture of Crops and Pastures

The simplification of vegetation and the disappearance of the litter layer under grassland and monocrop production systems lead to a decrease in faunal diversity. Although root systems (especially of grasses) can be extensive and explore vast areas of soil, the root exudates from one single crop will attract only a few different microbial species.

This in turn will affect the predator diversity. The more opportunistic pathogen species will be able to acquire space near the crop and cause harm. Continuous cultivation and grazing also leads to compaction of soil layers, which in turn affects the circulation of air. An aerobic conditions in the soil stimulate the growth of different micro-organisms, resulting in more pathogenic organisms.

High Harvest Index

One of the consequences of the green revolution was the replacement of indigenous varieties of species with high-yielding varieties (HYVs). These HYVs often produce more grain and less straw, compared with locally developed varieties; the harvest index of the crop (ratio of grain to total plant mass aboveground) is increased.

From a production point of view, this is a logical approach. However, this is less desirable from a conservation point of view. Reduced amounts of crop residues remain after harvest for soil cover and organic matter, or for grazing of livestock (which results in manure). Moreover, where animals graze the residues, even less remains for conservation purposes.

Decrease in Organic Matter Supply

Burning of Natural Vegetation and Crop Residues

The burning of maize, rice and other crop residues in the field is a common practice (Plate 5). Residues are usually burned to help control insects or diseases or to make fieldwork easier in the following season. Burning destroys the litter layer and so diminishes the amount of organic matter returned to the soil. The organisms that inhabit the surface soil and litter layer are also eliminated. For future decomposition to take place, energy has to be invested first in rebuilding the microbial community before plant nutrients can be released. Similarly, fallow lands and bush are burned before cultivation. This provides a rapid supply of P to stimulate seed germination. However, the associated loss of nutrients, organic matter and soil biological activity has severe long-term consequences.

Removal of Crop Residues

Many farmers remove residues from the field for use as animal feed and bedding or to make compost (Plate 7). Later, these residues return to contribute to soil fertility as manures or composts. However, residues are sometimes removed from the field and not returned. This removal of plant material impoverishes the soil as it is no longer possible to recycle the plant nutrients present in the residues.

Increased Decomposition Rates

Tillage Practices

Tillage is one of the major practices that reduces the organic matter level in the soil. Each time the soil is tilled, it is aerated. As the decomposition of organic matter and the liberation of C are aerobic processes, the oxygen stimulates or speeds up the action of soil microbes, which feed on organic matter.

This means that:

- When ploughed, the residues are incorporated in the soil together with air and come into contact with many micro-organisms, which accelerates the carbon cycle. The decomposition is faster, resulting in the formation of less stable humus and an increased liberation of CO_2 to the atmosphere, and thus a reduction in organic matter.

- The residues on the soil surface slow the carbon cycle because they are exposed to fewer micro-organisms and thus wane more slowly, resulting in the production of humus (which is more stable), and liberating less CO_2 to the atmosphere.

Table: *Tillage induced flush of decomposition of organic matter*

Type of tillage	*Organic matter lost in 19 days(kg/ha)*
Mouldboard plough + disc harrow (2x)	4 300
Mouldboard plough	2 230
Disc harrow	1 840
Chisel plough	1 720
Direct seeding	860

Source: Glanz, 1995

In terms of short-term organic matter loss, the more a soil is tilled, the more the organic matter is broken down. There are also longer-term losses, attributed to repeated, annual cultivation. Cropping systems that return little residue to the soil accelerate this decline.Many modern cropping systems combine frequent tillage with small amounts of residue, with resultant reductions in the organic matter content of many soils.

Historically, manure application (from farm livestock) was common, and it was a dynamic way of maintaining organic matter levels despite repeated cultivation and low residue returns to the soil.

Increased on-farm mechanization has reduced livestock numbers, so this source of organic material has been reduced considerably.

Organic matter production and conservation is affected dramatically by conventional tillage, which not only decreases soil organic matter but also increases the potential for erosion by wind and water (Plate 8). The impact occurs in many ways:

- Ploughing leaves no residues on the soil surface to lessen the impact of rain.
- Ploughing reduces the quantity of food sources for earthworms and disturbs their burrows and living space,

hence populations of certain species decrease drastically. Moreover, reduction of earthworm numbers reduces their impact, through burrowing, in increasing porosity and aeration (particularly continuous macropores) and lowers their ability to bury and incorporate plant residues, which facilitates rapid decomposition of organic matter.

- Tillage by repeated hoeing or discing smoothes the surface and destroys natural soil aggregates and channels that connect the surface with the subsoil, leaving the soil susceptible to erosion. Old root channels and earthworm holes are eliminated, as are the cracks between natural aggregates. The large pores, the ones destroyed by conventional tillage practices, are necessary to conduct water into the soil during rainfall.
- The development of a plough pan or hoe pan, a layer of compacted soil resulting from smearing action at the bottom of the plough or hoe, may retard both root penetration and water infiltration.
- Ploughing or discing under dry conditions exacerbates the pulverization of the soil, causing the soil surface to crust more easily, leading to greater water runoff and erosion. This is exacerbated by reduced soil surface roughness, which leaves few depressions for temporary storage of water during intense storms.
- Increased runoff during rainstorms may also increase the possibility of drought stress later in the season, because water that runs off the field does not infiltrate into the soil to remain available to plants.

In some circumstances, imbalances of certain soil organisms can disrupt soil structure and processes, e.g. certain earthworm species in rice fields or pastures.

Drainage

Decomposition of organic matter occurs more slowly in poorly aerated soils, where oxygen is limiting or absent, compared with well-aerated soils. For this reason, organic matter accumulates in wet soil environments. Soil drainage is determined strongly by topography - soils in depressions at the bottom of hills tend to remain wet for extended periods of time because they receive water (and sediments) from upslope. Soils may also have a layer in the subsoil that inhibits drainage, again exacerbating waterlogging and reduction in organic matter decomposition.

In a permanently waterlogged soil, one of the major structural parts of plants, lignin, does not decompose at all. The ultimate consequence of extremely wet or swampy conditions is the development of organic (peat or muck) soils, with organic matter contents of more than 30 percent. Where soils are drained artificially for agricultural or other uses, the soil organic matter decomposes rapidly.

Fertilizer and Pesticide Use

Initially, the use of fertilizer and pesticides enhances crop development and thus production of biomass (especially important on depleted soils). However, the use of some fertilizers, especially N fertilizers, and pesticides can boost micro-organism activity and thus decomposition of organic matter.

The chemicals provide the microorganisms with easy-to-use N components. This is especially important where the C: N ratio of the soil organic matter is high and thus decomposition is slowed by a lack of N.

Chapter 8

Some Diseases of the Soil

Soil Erosion

Perhaps the most widespread and the most important disease of the soil at the present time is soil erosion, a phase of infertility to which great attention is now being paid. Soil erosion in the very mild form of denudation has been in operation since the beginning of time. It is one of the normal operations of Nature going on everywhere. The minute mineral particles which result from the decay of rocks find their way sooner or later to the ocean, but many may linger on the way, often for centuries, in the form of one of the constituents of fertile fields. This phenomenon can be observed in any river valley. The fringes of the catchment area are frequently uncultivated hills, through the thin soils of which the underlying rocks protrude. These are constantly weathered and in the process yield a continuous supply of minute mineral fragments in all stages of decomposition.

The slow rotting of exposed rock surfaces is only one of the forms of decay: the surfaces not exposed are also subject to change. The covering of soil is no protection to these underlying strata, but rather the reverse, because the soil water, containing carbon dioxide in solution, is constantly disintegrating the parent rock, first producing subsoil and then actual soil. In this way the constant supply of minerals — like phosphates, potash, and the trace elements needed by crops and livestock — are automatically transferred to the surface soil from the great mineral reservoir of the primary and secondary rocks. Simultaneously with these disintegration processes the normal decay of animal and vegetable remains on the surface of the soil is giving rise to the formation of humus.

All these processes combine to start up denudation. The fine soil particles of mineral origin, often mixed with fragments of humus, are gradually removed by rain, wind, snow, or ice to lower regions. Ultimately the rich valley lands are reached, where the accumulations may be many feet in thickness. One of the main duties of the streams and rivers which drain the valley is to transport these soil particles into the sea, where fresh land can be laid down. The process looked at as a whole is nothing more than Nature's method of the rotation, not of the crop, but of the soil itself. When the time comes for the new land to be enclosed and brought into cultivation, agriculture is born again. Such operations are well seen in England in Holbeach Marsh and similar areas round the Wash. From the time of the Romans to the present day new areas of fertile soil, which now fetch £100 an acre or even more, have been recreated from the uplands by the Welland, the Nene, and the Ouse. All this fertile land, perhaps the most valuable in England, is the result of two of the most widespread processes in Nature — weathering and denudation.

But Nature has devised a most effective brake. The nature of this retarding mechanism is of supreme importance, because it provides the key to the solution of the problem of soil erosion. Nature's control of the rate of denudation is to create the compound soil particle. The fragments of mineral matter derived from the weathering of rocks are combined by means of the specks of glue-like organic matter supplied mostly by the dead bodies of the soil bacteria which live on humus; as in a building made of bricks, some suitable cementing material is needed before the fragments of mineral matter in the soil can cohere. There must be sufficient of this cement of the right type always ready, so that when the mineral fragments come together a piece of glue is there at hand of a size corresponding to the minute areas of contact. This involves the constant production of large quantities of this bacterial cement. Provided, however, that we keep up the bacterial population of the land in any catchment area, the supplies of glue for making new compound soil particles and for repairing the old ones will be assured.

It will be seen from this how fundamentally important is the role of humus. It is the humus which feeds the bacterial life, which, so to say, glues the soil together and makes it effective. If the supply of glue is allowed to fall into arrears, the compound soil particles will soon lie about in ruins and so provide more raw material for speeding up the process of denudation. The mineral particles are thereby released and ready for their final journey by water to the sea to form new soil, or by wind to form a new dust bowl and so begin a new desert.

It is when the tempo of denudation is vastly accelerated by human agencies that a perfectly harmless natural process becomes transformed into a definite disease of the soil. The condition known as soil erosion — a man-made disease— is then established. *It is, however, always preceded by infertility:* the inefficient, overworked, dying soil is at once removed by the operations of Nature and hustled towards the ocean, so that new land can be created and the rugged individualists — the bandits of agriculture — whose cursed thirst for profit is at the root of the mischief can be given a second chance. Nature is anxious to make a new and better start and naturally has no patience with the inefficient. Perhaps when the time comes for a new essay in farming, mankind will have learnt the great lesson — how to subordinate the profit motive to the sacred duty of handing over unimpaired to the next generation the heritage of a fertile soil. Soil erosion is nothing less than the outward and visible sign of the complete failure of a farming policy. The root causes of this failure are to be found in ourselves.

The damage already done by soil erosion all over the world, looked at in the mass, is very great and is rapidly increasing. The regional contributions to this destruction, however, vary widely. In some areas like north-western Europe, where most of the agricultural land is under a permanent or temporary cover crop (in the shape of grass or leys) and there is still a large area of woodland and forest, soil erosion is a minor factor in agriculture. In other regions like parts of North America, Africa, Australia, New Zealand, and the countries bordering the Mediterranean, where extensive deforestation has been practiced and where almost uninterrupted cultivation has been the rule, large tracts of land once fertile have been almost completely destroyed. The United States of America is perhaps the only country where anything in the nature of an accurate estimate of the damage done by erosion has been made. Theodore Roosevelt first warned the country as to its national importance. Then came the Great War with its high prices, which encouraged the wasteful exploitation of soil fertility on an unprecedented scale. A period of financial depression, a series of droughts and dust storms, emphasized the urgency of the salvage of agriculture.

During Franklin Roosevelt's presidency soil conservation became a political and social problem of the first importance. In 1937 the condition and needs of the agricultural land of the United States of America were appraised. No less than 253,000,000 acres, or 61 per cent of the total area under crops, had either been completely or partly destroyed or had lost most of its fertility. Only 161,000,000 acres, or 39 per cent of the cultivated area, could be safely farmed by present methods. In less than a century the United States has, therefore, lost nearly three-fifths of its agricultural capital. If the whole of the potential resources of the country could be utilized and the best possible practices introduced everywhere, about 447,466,000 acres could be brought into use — an area actually greater than the present crop land of 415,334,931 acres. The position, therefore, is not hopeless. It will, however, be very difficult, very expensive, and very time-consuming to restore the vast areas of eroded land even if money is no object and large amounts of manure are used and green-manure crops are ploughed under.

Such, in this great country, are the results of misuse of the land. The causes of this misuse include lack of individual knowledge of soil fertility on the part of the pioneers and their descendants; the traditional attitude which regarded the land as a source of profit; defects in farming systems, in tenancy, and finance — most mortgages contain no provisions for the maintenance of fertility; instability of agricultural production as carried out by millions of individuals, prices, and income, in contrast to industrial production carried on by a few large corporations. The need for maintaining a correct relation between industrial and agricultural production, so that both can develop in full swing on the basis of abundance, has only recently been understood. The country was so vast, its agricultural resources were so immense, that the profit seekers could operate undisturbed until soil fertility — the country's capital — began to vanish at an alarming rate.

The resources of the Government are now being called up to put the land in order. The magnitude of the effort, the mobilization of all available knowledge, the practical steps that are being taken to save what is left of the soil of the country and to help Nature to repair the damage already done are graphically set out in *Soils and Men,* the Year Book of the United States Department of Agriculture of 1938. This is perhaps the best local account of soil erosion which has yet appeared. The progress that has been made in recent years can be followed in *Soil Conservation,* a monthly periodical issued by the Soil Conservation Service of the United States Department of Agriculture, Washington, D.C.

The rapid exploitation of Africa was soon followed by soil erosion. In South Africa, a pastoral country, some of the best grazing areas are already semi-desert. The Orange Free State in 1879 was covered with rich grass, interspersed with reedy pools, where now only useless gullies are found. Towards the end of the nineteenth century, it began to be realized all over South Africa that serious overstocking was taking place. In 1928 the Drought Investigation Commission reported that soil erosion was extending rapidly over many parts of the Union and that the eroded material was silting up reservoirs and rivers and causing a marked decrease in the underground water supplies.

The cause of erosion was considered to be the reduction of vegetal cover brought about by incorrect veldt management — the concentration of stock in kraals, overstocking, and indiscriminate burning to obtain fresh autumn or winter grazing. In Basutoland, a normally well watered country, soil erosion is now the most immediately pressing administrative problem. The pressure of population has brought large areas under the plough and has intensified overstocking on the remaining pasture. In Kenya the soil erosion problem has become serious during the last ten years, both in the native reserves and in the European areas. In the former, wealth depends on the possession of large flocks and herds; barter is carried on in terms of livestock; the bride price is almost universally paid in anima's s; numbers rather than quality are the rule.

The natural consequence is overstocking, over-grazing, and the destruction of the natural covering of the soil. Soil erosion is the inevitable result. In the European areas, erosion is caused by long and continuous over-cropping without the adoption of measures to prevent the loss of soil and to maintain the humus content. Locusts have of late been responsible for greatly accelerated erosion; examples are to be seen when the combined effect of locusts and goats has resulted in the loss of a foot of surface soil in a single rainy season.

The countries bordering the Mediterranean provide striking examples of soil erosion, accompanied by the formation of deserts which are considered to be due to one main cause — the slow and continuous deforestation of the last 3,000 years. Originally well wooded, no forests are to be found in the Mediterranean region proper. Most of the original soil has been washed away by the sudden winter torrents. In North Africa the fertile cornfields which existed in Roman times are now desert. Ferrari in his book on woods and pastures refers to the changes in the soil and climate of Persia after its numerous and majestic parks were destroyed; the soil was transformed into sand; the climate became arid and suffocating; springs first decreased and then disappeared. Similar changes took place in Egypt when the forests were devastated; a decrease in rainfall and in soil fertility was accompanied by loss of uniformity in the climate. Palestine was once covered with valuable forests and fertile pastures and possessed a cool and moderate climate; today its mountains are denuded, its rivers are almost dry, and crop production is reduced to a minimum.

The above examples indicate the wide extent of soil erosion, the very serious damage that is being done, and the fundamental cause of the trouble — misuse of the land, resulting in the destruction of the compound soil particles. In dealing with the remedies which have been suggested and which are now being tried out, it is essential to envisage the real nature of the problem. It is nothing less than the repair of Nature's drainage system — the river — and of Nature's method of providing the countryside with a regular water supply. The catchment area of the river is the natural unit in erosion control. In devising this control we must restore the efficiency of the catchment area as a drain and also as a natural storage of water. Once this is accomplished, we shall hear very little about soil erosion. Japan provides perhaps the best example of the control of soil erosion in a country with torrential rains, highly erodible soils, and a topography which renders the retention of the soil on steep slopes very difficult. Here erosion has been effectively held in check by methods adopted regardless of cost, for the reason that the alternative to their execution would be national disaster.

The great danger from soil erosion in Japan is the deposition of soil debris from the steep mountain slopes on the rice fields below. The texture of the rice soils must be maintained so that the fields will hold water and allow of the minimum of through drainage. If such areas become covered with a deep layer of permeable soil, brought down by erosion from the hillsides, they would no longer hold water and rice cultivation — the mainstay of Japan's food supply — would be out of the question. For this reason the country has spent as much as ten times the capital value of eroding land on soil conservation work, mainly as an insurance for saving the valuable rice lands below. Thus, in 1925 the Tokyo Forestry Board spent 453 yen (£45) per acre in anti-erosion measures on a forest area valued at 40 yen per acre in order to save rice fields lower down valued at 240 to 300 yen per acre.

The dangers from erosion have been recognized in Japan for centuries and an exemplary technique has been developed for preventing them. It is now a definite part of national policy to maintain the upper regions of each catchment area under forest as the most economical and effective method of controlling flood waters and insuring the production of rice in the valleys. For many years erosion control measures have formed an important item in the national budget. According to Lowdermilk, erosion control in Japan is like a game of chess. The forest engineer, after studying his eroding valley, makes his first move, locating and building one or more check dams. He waits to see what Nature's response is. This determines the next move which may be another dam or two, an increase in the former dam, or the construction of retaining side walls. After another pause for observation a further move is made and so on until erosion is checkmated. The operation of natural forces, such as sedimentation and re-vegetation, are guided and used to the best advantage to keep down costs and to obtain practical results. *No more is attempted than Nature has already done in the region.* By 1929 nearly 2,000,000 hectares of protection forests were used in erosion control. These forest areas do more than control erosion. They help the soil to absorb and retain large volumes of rain water and to release it slowly to the rivers and springs.

China, on the other hand, presents a very striking example of the evils which result from the inability of the administration to deal with the whole of a great drainage area as one unit. On the slopes of the upper reaches of the Yellow River extensive soil erosion is constantly going on. Every year the river transports over 2,000,000,000 tons of soil, sufficient to raise an area of 400 square miles by five feet. This is provided by the easily erodible loess soils of the upper reaches of the catchment area. Some of the mud is deposited in the river bed lower down, so that the embankments which contain the stream have constantly to be raised. Periodically the great river wins in this unequal contest and destructive inundations result. The labour expended on the embankments is lost, because the nature of the erosion problem as a whole has not been grasped, and the area drained by the Yellow River has not been studied and dealt with as a single organism. The difficulty now is the over-popuration of the upper reaches of the catchment area, which prevents afforestation and laying down to grass. Had the Chinese maintained effective control of the upper reaches — the real cause of the trouble — the erosion problem in all probability would have been solved long ago at a lesser cost in labour than that which has been devoted to the embankment of the river. China, unfortunately, does not stand alone in this matter. A number of other rivers, like the Mississippi, are suffering from overwork, followed by periodical floods as the result of the growth of soil erosion in the upper reaches.

Although the damage done by uncontrolled erosion all over the world is very great and the case for action needs no argument, nevertheless there is one factor on the credit side which has been overlooked. A considerable amount of new soil is being constantly produced by natural weathering agencies from the subsoil and the parent rock. This, when suitably conserved, will soon re-create large stretches of valuable land.

One of the best regions for the study of this question is the black cotton soil of Central India which overlies the basalt. Here, although erosion is continuous, the soil does not often disappear altogether, for the reason that, as the upper layers are removed by rain, fresh soil is re-formed from below. The large amount of earth so produced is well seen in the Gwalior State, where the late ruler employed an irrigation officer, lent by the Government of India, to construct a number of embankments, each furnished with spillways, across many of the valleys, which had suffered so badly by uncontrolled rain wash in the past that they appeared to have no soil at all, the scrub vegetation just managing to survive in the crevices of the bare rock. How great is the annual formation of new soil, even in such unpromising circumstances, must be seen to be believed. In a few years the construction of embankments was followed by stretches of fertile land which soon carried fine crops of wheat. A brief illustrated account of the work done by the late Maharaja of Gwalior would be of great value at the moment for introducing a much needed note of optimism in the consideration of this soil erosion problem. Things are not quite so hopeless as they are often made to appear.

Why is the forest such an effective agent in the prevention of soil erosion? The forest does two things: (1) the trees and undergrowth break up the rainfall into fine spray and the litter on the ground further protects the soil from the impact of the descending water stream; (2) the residues of the trees and animal life met with in all woodlands are converted into humus, which is then absorbed by the soil underneath, increasing its porosity and water-holding power; the soil cover and the soil humus together prevent erosion and at the same time store large volumes of water. These factors — soil protection, soil porosity, and water retention — conferred by the living forest cover, provide the key to the solution of the soil erosion problem. All other purely mechanical remedies, such as terracing and drainage, are secondary matters, although, of course, important in their proper place.

The secret of soil conservation is thus seen to lie, first, in maintaining the soil cover in good condition to ensure that the rainfall is received on the surface in a proper manner with no disturbance of the soil below, and second, in conserving ample supplies of humus so that by means of the compound soil particles the water, when it has descended, is adequately absorbed and stored: as well might we expect a living creature to survive without its protective skin as to suppose that the earth can live without her proper covering. The forest has been cited as the pre-eminent example of these protective devices, for the leafage is thick and the ground litter abundant. In the absence of forest some form of grass cover is the natural protective agent which will for centuries often maintain the soil in good heart. Indeed, this device of the grass cover is far more efficient than might be supposed possible.

The accumulations of humus under a grass carpet are often immense; they are, indeed, so extraordinary that they can be described as veritable mines of fertility. This is proved by the fact that an agriculture based on their spoliation can, in favourable circumstances, continue for many years before it fades out. But fade out it must if the humus is never restored. Williams regarded grass as the basis of all agricultural land utilization and the soil's chief weapon against the plundering instincts of humanity.

He advanced the hypothesis that the decay of past civilizations was due to the wholesale ploughing up of grass necessitated by the increasing demands of civilization. His views are exerting a marked influence on soil conservation policy in the U.S.S.R. and indeed apply to many other countries.

Grass is a valuable factor in the correct design and construction of surface drains. Whenever possible these should be wide, very shallow, and completely grassed over. The runoff then drains away as a thin sheet of clear water, leaving all the soil particles behind. The grass is thereby automatically manured and yields abundant fodder. This simple device was put into practice at the Shahjahanpur Sugar Experiment Station in India. The earth service roads and paths were excavated so that the level was a few inches below that of the cultivated area. They were than grassed over, becoming very effective drains in the rainy season, carrying off the excess rainfall as clear water without any loss of soil.

If we regard erosion as the natural consequence of improper methods of agriculture and the catchment area of the river as the natural unit for the application of soil conservation methods, the various remedies available fall into their proper place. The upper reaches of each river system must be afforested; cover crops, including grass and leys, must be used to protect the arable surface whenever possible; the humus content of the soil must be increased and the crumb structure restored, so that each field can drink in its own rainfall; overstocking and overgrazing must be prevented; simple mechanical methods for conserving the soil and regulating the runoff, like terracing, contour cultivation and contour drains, must be utilized. There is, of course, no single anti-erosion device which can be universally adopted. The problem must, in the nature of things, be a local one. Nevertheless, certain guiding principles exist which apply everywhere. First and foremost is the restoration and maintenance of the crumb structure of the soil, so that each acre of the catchment area can do its duty by absorbing its share of the rainfall.

The Roman Occupation

At the time of the Roman invasion most of the island in which we are living was under forest or marsh: only a portion of the uplands was under grass or crops: the population was very small. After the conquest of the country the Romans began to develop it by the creation on the areas already cleared of an agricultural unit — new to Great Britain — known as the villa.

These villas were large farms under single ownership run by functionaries each responsible for a particular type of animal or crop and worked by slave labour. These units followed to some extent the methods of the *latifundia* of Italy and were designed for the production of food for the legions garrisoning the island and those stationed in Gaul. Wheat — an exhausting crop — was an important item in Roman agriculture, for the reason that this cereal provided the chief food *(frumentum)* of the soldiers. The extent of the export of grain to Gaul will be evident from the fact that in the reign of the Emperor Julian no less than 800 wheat ships were sent from Britain to the Continent. The exhaustion of the soils of the island began even before the Roman occupation. The heavy soil-inverting mould board plough, which invariably wears out the land, was already in use when the Romans arrived, and was probably brought by the Belgic tribes who conquered and settled in the south-eastern part of the country. They lived in farmsteads and cultivated large open fields. They were highly skilled agriculturists and exported to Gaul a considerable quantity of their main product — wheat. This practice was developed by the Roman villas which followed and in this way the slow exhaustion of the lighter soils of the downlands of the southeast became inevitable. After an occupation which lasted some 400 years and which contributed little or nothing of permanent value to the agriculture of the island beyond some well-designed roads, the legions evacuated the island and left the Romanized population to look after itself. This they failed to do: the country was soon conquered by the Saxon invaders, in the course of which much destruction of life and property took place. One result was the creation of a new type of farming.

The Saxon Conquest

The settlement of Nordic people in our island is the governing event both of British history and of British agriculture. The new settlers had inhabited the belts of land around the Weser and the Elbe and their first contact with Britain was as raiders; their operations were in the nature of reconnaissance to ascertain the chances of settlement. The Anglo-Saxon migration to Britain was a colonization preceded by conquest, in which the farming system of the Romanized population was, in the midland area at any rate, destroyed. In the east, southeast, and western portions of the island some relics of Roman and Celtic methods survived.

Our forefathers brought with them from the opposite shores of the North Sea their wives, children, livestock, and a complete fabric of village life. The immigrants, being country folk, wanted to live in rural huts with their cattle round them and their land nearby, as they did in Germany. The numerous villages they formed reproduced in all essentials those they had left behind on the mainland. Our true English villages are, therefore, not Celtic, are not Roman, but purely and typically German. The Roman villas were replaced by a new system of farming — the Saxon manor — in which the tenants held land in return for service. The lord and his retainers shared the land, each bound to perform certain duties determined by custom. The manors took centuries to evolve. By A.D. 800 they had developed into a permanent system which provided the material for the Domesday Book of the Normans, by which taxation was assessed and a rigid feudal system became firmly established.

The Open-Field System

The first general feature that strikes us in early Anglo-Saxon England is the strip cultivation of the arable land on the open-field system. This system was a communal agricultural institution started by people who had to get a living out of the soil. They had progressed as far as to use the plough and had a common fund of experience. Everyone pursued the same system of farming. The arrangement of the open fields was, however, by no means uniform. No fewer than three distinct types arose, corresponding to as many different influences exerted by people who had early occupied the country. The large central midland area, stretching from Durham to the Channel and from Cambridgeshire to Wales, is the region where Germanic usage prevailed.

The southeast was characterized by the persistence of Roman influence, a circumstance which implies that the conquest was less destructive there than in the north and west. The counties of the south-west, northwest, and the north retained Celtic agrarian usages in one form or another, which is easily understood in view of the difficulty with which, as we know, these districts were slowly overpowered by the invaders. The midland area was thus the region where the Anglo-Saxons were most firmly established and where the subjugation of the fifth century was most thorough.

The Romano-Celtic people who remained were not numerous enough to preserve any traces of Roman or Celtic methods of tilling the soil. Throughout this extensive region a two-field and a three-field system, or sometimes a mixture of the two, prevailed. This field arrangement was a custom prevalent in Germany, especially east and south of the Weser. The chief characteristic of the two- and three-field type of tillage was the distribution of the parcels of arable land (which made up the holdings of the customary tenants) equally amongst the two or three fields. The cropping was so arranged that one field in the two-field system and two fields in the three-field system were cropped every year, and thus one-half or one-third of the township's arable land lay fallow and was used for common grazing — a point which is always emphasized in the midland system. Besides the cultivated open fields, for which the best land was always used, the village lands consisted of grassland for mowing on the wetter parts, and commons or woodlands on the poorer parts. Ploughing was the all-important operation of medieval tillage and was carried out on a co-operative basis, and demanded a team of eight draught animals yoked to a heavy plough. This, of course, was beyond the reach of any but the largest and most prosperous tenants. Communal ploughing in Saxon times was, therefore, inevitable. It was the difficulty of replacing this communal ploughing that delayed agricultural progress in many parts of the country.

The open-field system repeated itself for centuries, not only in England but in a great part of Europe — nations living under very different conditions, in very different climates, and on very different soils adopted the open-field system again and again without having borrowed it from each other. This could not but proceed from some pressing necessity. The open-field system is communal in its very essence. Every trait which makes it strange and inconvenient from the point of view of individualistic interests renders it highly appropriate to a state of things ruled by communal conceptions — right of common usage — communal arrangements of ways and time of cultivation. These are the main features of open-field husbandry and all point to one origin — the formation in early Anglo-Saxon society of a village community of shareholders of free and independent growth. It must be borne in mind that the open-field prevailed during the period of national formation of the English people and its influence on the life of the village community must have been very great. The sense of personal responsibility, which the system of communal work created, made it a vital factor in the social education of the people.

The Depreciation of Soil Fertility

Open-field farming is, as a rule, balanced: the fertility used up in growth is made good before the next crop is sown. Compared with our modern standards, however, the yield is remarkably low and the removal of fertility by such small crops is made up for by the recuperative processes operating in the soil (non-symbiotic fixation of nitrogen and so forth). The surplus of available humus originally left by the forest is depleted at an early stage and an equilibrium is established, the yield adjusting itself to the amount of fertility added each year by natural processes, this in its turn is influenced by climate and methods of cultivation. For example, in the peasant cultivation of northwest India at the present day a perfect balance has been established between losses and gains of fertility. The village land on which corn crops are grown has been cultivated for upwards of 2,000 years without manure beyond the droppings of the livestock during the fallow period between harvest and the rains. But the Indian cultivators use primitive scratch ploughs and are most careful not to draw on the reserves of organic material in the soil, as its texture depends on this. They produce crops entirely on the current account provided by the annual increments of fertility. The yield has settled down to 8 maunds (658 lb. per acre) of wheat on unirrigated land, and 12 maunds (987 lb.) of wheat on irrigated land, and this yield has been constant for many centuries.

The same processes were operating in the English open fields. The reserve of humus in the soils originally under forest, which the Saxons brought into cultivation, was soon used up and the yield was determined by the annual additions of fertility to the soil by natural means. But in our cold and sunless climate and on our ill-drained, poorly aerated soils this is far less than in the semi-tropical conditions of northern India. Moreover, and this point must be stressed, the Saxons from the earliest times used a soil-inverting plough, which has a marked tendency to exhaust the humus in the soil if provision is not made for the regular supply of sufficient farmyard manure. In fact, recent experience in many parts of the world is proving that the continued use of heavy soil-inverting, tractor-driven implements, without sufficient farmyard manure to manure the land, promptly leads to catastrophic consequences.

The first recorded references to the mould board plough speak of it in Gaul, but some authorities quoted by Vinogradoff (*The Growth of the Manor*) suggest that it was borrowed by the Germanic people from the Slavs, and in view of the soil types found in Slav territory this may easily be so. The evolution of the big plough was due to soil requirements as settled agricultural life developed in the heavy, moist soils of north Europe after the forests had been cleared. The mould board plough determined the lay-out of the open fields. It divided the arable areas into a succession of lands. It needed a headland to turn on, and there was a limit to the length of furrow a team of oxen could plough before needing the relief got by stopping and turning.

This furrow- long or furlong became one of our units of length. It was usual to keep the land in high ridges running along the slopes to facilitate surface drainage, an important point in England. The ridges varied in width according to the nature of the soil. In very heavy clays they were sometimes no more than three yards wide. In lighter soils they might be twenty-two yards wide. These ridges may be seen in many places today on grassland which was under the plough in earlier centuries. From this brief description it will be seen that the open fields cultivated with the heavy medieval plough were laid out in strips. The main feature of the heavy mould board plough was its high penetrating power, and it could be used on the heavier types of soil where the light scratch plough of the Celts and Italians would be useless. It thus enabled the cropped area in England to be greatly extended by the cultivation of the heavy soil of the valleys and plains which first had to be slowly carved out of the forest. It owed its superiority to an iron share, a courter, and a wooden mould board so suitable on wet land. This primitive implement gave us the plough as we know it today. The principle of our modern plough is identical and, except for the fact that it is now made entirely of iron, it is almost the same in detail.

The open-field system of the Middle Ages was bound to fail because it involved burning the candle at both ends and also in the middle. First the natural recuperation processes in the soil were hampered by low temperatures and poor soil aeration; second, such supplies of farmyard manure as were available were by custom mostly bestowed on the lord's demesne lands, and besides were inadequate because only a portion of the livestock could be wintered; finally the soil-inverting plough led to the oxidation of the stores of soil humus faster than it could be recreated and was bound to wear out the land.

The Low Yield of Wheat

The failure of the open-field system is proved by the low yield of wheat. All authorities agree that the yield of wheat in England during the Middle Ages was at a very low level, though it does not appear to have varied greatly.

It may be noted that there was never any question of complete exhaustion of the wheat-growing land, such as occurred in Mesopotamia and in the Roman wheat-growing regions of North Africa, where the soil, owing to over-cropping and in some instances to over-irrigation aggravated by special climatic conditions, became sterile and was transformed into desert.

This could not so easily happen in the moist, temperate climate of Great Britain. What happened in the Middle Ages in England was that the yield of corn was not high enough for the requirements of the growing social and economic life of the country.

The material for a quantitative estimate of wheat yields in this period is necessarily very scanty, but in the case of some large estates records are available for a considerable period of years of the seed sown in one year and the grain threshed in the following year, and these form the basis of the best estimates of medieval yields. Sir William Beveridge, using this method, investigated the yield of wheat for the years 1200 to 1450 on eight manors, including that of Wargrave, situated in seven different counties belonging to the Bishop of Winchester.

The average yield per acre was 1.17 quarters or 9.36 measured bushels, equivalent to 7.48 bushels of 60 lb. It is to be noted that these estimates were all from demesne lands which were probably better cultivated and better manured than the land of the customary tenants. Other authorities confirm these figures.

The figures of yield given above help to account for the changes which marked the end of the Middle Ages. The amount of food was becoming insufficient for the growing population. But another factor was steadily developing, which finally assumed the dimensions of an avalanche and led to the reform of manorial farming. This was disease, a matter which must now be discussed.

The Black Death

That the agriculture of the Middle Ages was unable to keep the population in health was first indicated by the frequent indications of rural unrest. But these were soon followed by the writing on the wall in the shape of the Black Death in 1348-9. This outbreak had been preceded by several years of dearth and pestilence, and it was succeeded by four visitations of similar disease before the end of the century. During its ravages it destroyed from one-third to one-half of the population. This seriously affected the labour supply, which was no longer sufficient to carry on the traditional methods of manorial farming, already beginning to be undermined by the growing tendency to replace service by money payments.

Land which could no longer be ploughed had to be laid down to grass and used for feeding sheep to produce more of the wool so urgently needed in Flanders and Lombardy. For the new farming the countryside had to be enclosed: first the lord's demesne and then the area under open fields began to be laid down to grass. The earth's green carpet not only fed the sheep, but gave the land a long rest: large reserves of humus were gradually built up under the turf: the fertility of the soil, which had been imperceptibly worn out by the mould board plough and the constant cropping of the manorial system, was gradually restored.

After a long period of rest of a century the land no longer returned only seven and a half bushels to the acre. The figures given above for the years 1200 to 1450 may be contrasted with the figures from a farm at Wargrave from 1612-20: in these years the average was 25.6 bushels of 60 lb. per acre. In the latter part of the sixteenth century the general average was eighteen bushels to the acre and even more. That this significant change was due to the restoration of soil fertility by humus formation under the turf there can be no doubt. It is more than probable that the slow regeneration of the soils of this country, which began after the Black Death, produced other results besides the improvement of crops and livestock. What of the effect of the produce of land in good heart on the most important crop of all — men and women? Were the outstanding achievements of the Tudor period one of the natural consequences of a restored agriculture? It may well be so.

Enclosure

When increasing population led once more to the breaking up of the grassland and the farmer returned to tillage, the land, after its long rest of upwards of a century, was again capable of responding to the demands made upon it. One result of this experience was an increased interest in enclosure. Instinct was leading to a search for an economic arrangement which would prevent soil exhaustion from being repeated in succeeding ages. Enclosed farms offered a solution, as they gave the farmer the chance of keeping his land in good condition by individual management in place of the easy-going farming of the open fields of old English village agriculture. They also offered to the enclosed farmer the opportunity of composting his straw in his cattle yards and producing as much farmyard manure as possible. This, in most cases, he did, and the plan succeeded. Nevertheless, the ancient open-field tillage husbandry had in its favour the authority of long tradition — a potent force with a suspicious and conservative peasantry. The peasant asked himself: In the case of a readjustment of holdings would not the strong profit and the weak suffer? There grew up a popular prejudice against enclosure and the improvement of the common fields, but in the end, after some centuries of contest, enclosure won.

The form which the enclosure movement took before it was completed was due to the peculiar form of government which came in with the English Revolution of 1688. By that event the landed gentry became supreme. The national and local administration was entirely in their hands, and land, being the foundation of social and political influence, was eagerly sought by them. They not unnaturally wished to direct the enclosure movement into channels which were in the interests of their estates. But in doing so they made some of the most outstanding contributions to farming ever made in our history.

The restoration of soil fertility which resulted from enclosure had a profound influence on both livestock and crops. The provision of more and better forage and fodder which followed the cultivation of clover and artificial grasses, coupled with the popularization of the turnip crop by Townshend in 1730, opened the door for the continuous improvement of livestock by pioneers like Bakewell. The result was that our livestock improved in size and in the quality of the meat. Between 1710 and 1795 the weights of cattle sold at Smithfield more than doubled. By 1795 beeves weighed 800 lb. as compared with 370 lb.; sheep went up from 28 lb. to 80 lb. The improvement in the yield of cereals was no less significant. That of rye or wheat rose from 6-8 bushels to the acre in the Middle Ages to 15-20 bushels; barley yielded up to 36 bushels, oats 32-40 bushels. All this was due to more and better food for the livestock and more manure for the land. More manure raised larger crops: larger crops supported much bigger flocks and herds.

Another change in the countryside accompanied the enclosures. The forests, which since Saxon times had been gradually cleared and converted into manorial lands, had by this process become exhausted. After the Civil War it was realized that the country was running short of the hardwoods needed for maintaining the fleet and for buildings and so forth. An era of tree planting, which continued for two hundred years, was inaugurated by the publication of Evelyn's *Sylva* in 1678. It was during this period that the English landscape as we know it today was created by the judicious laying out of parks, artificial lakes, groups of trees, and woods. All this planting provided an important factor in the maintenance of soil fertility. The roots of the trees and the hedges combed the subsoil for minerals, embodied these in the fallen leaves and other wastes of the trees and shrubs, and so helped to maintain the humus in the soil, as well as the circulation of minerals. The roots also acted as subsoil ploughs and aerating agencies. The cumulative effect of the trees and hedges, which accompanied enclosure, in maintaining soil fertility has passed almost unnoticed. Nevertheless, its importance in humus production and in the availability of minerals must be considerable.

While the policy of enclosure, combined with tree-planting and the creation of the existing English landscape, arrested the fall in soil fertility which was inherent in the open-field system, the freedom of action which followed enclosure afforded full scope to the improver. The restoration of British agriculture owes much to the pioneers among the landlords themselves, particularly to Coke of Holkham (1776-1816), who did much to introduce the Norfolk four-course system — (1) turnips, (2) barley, (3) seeds (clover and rye grass), (4) wheat — into general practice and so to achieve at long last an approach to Nature's law of return. Besides his championship of the Norfolk four-course system, his achievements include the conversion of 2,000,000 acres of waste into well-farmed and productive land, the prevention of famine in England during the Napoleonic Wars, the solution of the rural labour problem in his locality by means of a fertile soil, the demonstration of the principle that money well laid out in land improvement is an excellent investment. He invested half a million sterling in his own property and thereby raised the rent roll of his estate from £2,200 a year to £20,000. He transformed agriculture in this country by the simple process of first writing his message on the land and then, by means of his famous sheep-shearing meetings, bringing it to the notice of the farming community.

But the replacement of the manorial system by individual farming in fenced fields was attended by some grave disadvantages. The large profits obtained from the sale of wool, for example, while they enriched the few, led to a new conception of agriculture. The profit motive began to rule the farmer; farming ceased to be a way of life and soon became a means of enrichment. Enterprising individuals were afforded considerable scope for using their farms to make money. At the same time, large numbers of less fortunate individuals deprived of their land had either to work for wages or seek a living in the towns.

The various Enclosure Acts, which covered a period of more than 600 years, 1235-1845, therefore led to a new agriculture, the enthronement of the profit motive in the national life, and to the exploitation of coal, iron, and minerals, which is customarily referred to as the Industrial Revolution. This arose from the activities of the tradesmen of the manor, whose calling was destroyed by the Enclosure Acts. The last of the Enclosure Acts, which finally put an end to the strip system of the open fields, was passed in 1845. About the same time the celebrated Broadbalk wheat plots of the Rothamsted Experimental Station were laid out. This field is divided into permanent parallel strips and cultivated on even more rigid lines than anything to be found in the annals of manorial farming. These plots never enjoy the droppings of livestock: till recently they never had the benefit of the annual rest provided by a fallow. Practically every agricultural experiment station all over the world has copied Rothamsted and adopted the strip system of cultivation. How can such experiments, based on an obsolete method of farming, ever hope to give a safe lead to practice? How can the higher mathematics and the ablest statistician overcome such a fundamental blunder in the original planning of these trials?

The strip system has also been adopted for the allotments round our towns and cities without any provision whatsoever on the part of the authorities to maintain the land in good heart by such obvious and simple expedients as subsoiling, followed by a rest under grass grazed by sheep or cattle, ploughing up, and sheet-composting the vegetable residues. Land under allotments should not be under vegetables for more than five years at a time; this should be followed by a similar period under grass and livestock.

The Industrial Revolution and Soil Fertility

The released initiative which accompanied the collapse of the manorial system was by no means confined to the restoration of soil fertility and the development of the countryside. The dispossessed craftsmen started all kinds of industries, in which they used as labour-saving devices first water power, then the steam engine, the internal combustion engine, and finally electrical energy. By these agencies the Industrial Revolution, which continues till this day, was set in motion. It has influenced farming in many directions. In the first place, industries have encroached on and seriously reduced the area under cultivation. But by far the most important demand of the Industrial Revolution was the creation of two new hungers — the hunger of a rapidly increasing urban population and the hunger of its machines. Both needed the things raised on the land: both have seriously depleted the reserves of fertility in our soils. Neither of these hungers has been accompanied by the return of the respective wastes to the land. Instead, vast sums of money were spent in completely sidetracking these wastes and preventing their return to the land which so sadly needed them. Much ingenuity was devoted to developing an effective method of removing the human wastes to the rivers and seas. These finally took the shape of our present-day water-borne sewage system. The contents of the dustbins of house and factory first found their way into huge dumps and then into incinerators or into refuse tips sealed by a thin covering of cinders or soil.

At first the additional demands for food and raw materials were met by the restored agriculture and the periodical ploughing up of grass. One of these demands was the vast quantities of corn needed to feed the urban population. The price of wheat was regulated for more than 150 years by a series of Corn Laws, which attempted to hold the balance between the claims of the farmers who produced the grain and those of the consumers and the industrialists who advocated cheap food for their workers, so that they could export their produce at a profit. But as the urban population expanded, the pressure on the fertility of the soil increased until, in 1845, a disastrous harvest and the potato famine compelled the Government in 1846 to yield. The "rain rained away" the Corn Laws (Prothero). Deprived of protection, farmers were forced to adopt new methods and to farm intensively. Many developments in farming occurred. Particular attention was paid to drainage: the first drain pipe was made in 1843; two years later the pipes were turned out by a machine. Liebig's famous essay in 1843 drew attention to the importance of manures, While better farm buildings and the preparation of better farmyard manure were adopted, two fatal mistakes were made. Artificial manures like nitrate of soda and superphosphate came into use: imported feeding stuffs for livestock began to take the place of home-grown food. British farming, in adopting these two expedients, because they appeared for the moment to be profitable, laid the foundations of much future trouble But in the use of better implements for the land and the provision of improved transport facilities the countryside was on firmer ground. The result of all these and other developments was a period of great prosperity for farming which lasted till late in the seventies of the last century.

The Great Depression of 1879

Then the blow fell. The year 1879, which I remember so vividly, was one of the wettest and coldest on record. The average yield of wheat fell to about fifteen bushels to the acre:

large numbers of sheep and cattle were destroyed by disease: the price of wheat fell to an undreamt-of level as the result of large importations from the virgin lands of the New World. The great depression of 1879 not only ruined many farmers, but it dealt the industry a mortal blow. Farmers were compelled to meet a new set of conditions — impossible from the point of view of the maintenance of soil fertility — which have been more or less the rule till the Great War of 1914-18 and the World War which began in 1939 provided a temporary alleviation as far as the sale of produce and satisfactory prices were concerned.

Since 1879 the standard of real farming in this country has steadily fallen. The labour force, particularly the supply of men with experience of and sympathy with livestock, markedly diminished and deteriorated in quality. Rural housing left much to be desired. Drainage was sadly neglected. The small hill farms, which are essential for producing cattle possessing real bone and stamina, fell on evil days. Our flocks of folded sheep, so essential for the upkeep of downland, dwindled. Diseases like foot-and-mouth, tuberculosis, mastitis, and contagious abortion became rampant. Less and less attention was paid to the care of the manure heap and to the maintenance of the humus content of the soil. The Shortcomings of Present-day Agricultural Research) replaced the muck mentality of our fathers and grandfathers. Murdered bread, deprived of the essential germ, replaced the real bread of the last century and seriously lowered the efficiency of our rural population. The general wellbeing of our flocks and herds fell far below that of some of our overseas competitors like the Argentine. But in this dark picture some rays of light could be detected. The pioneers were busy demonstrating important advances. Among these two are outstanding: (1) the Clifton Park system of farming based on deep-rooting plants in the grass carpet, and (2) the use of the subsoiler for breaking up pans under arable and grass, and so preparing the ground for another great advance — the mechanized organic farming of tomorrow.

The Second World War

Such, generally speaking, was the condition of British agriculture in September 1939, when the second world war began and the submarine menace for the second time brought national starvation into the picture. What an opportunity was provided for a Coke of Norfolk for making use of a portion of the resources of a great nation to set British farming on its feet for all time by the simple expedient of restoring and maintaining soil fertility!

What an opening was given to the pioneers of human nutrition and the apostles of preventive medicine for feeding the men and women defending the country on the fresh produce of fertile soil and so initiating the greatest food reform in our history! But the potential Cokes of Norfolk had been liquidated or discouraged by many years of death duties, which had destroyed most of our agricultural capital and deprived the countryside of its natural leaders who, in years gone by, had done so much for farming. The apostles of real nutrition and of preventive medicine, such as the panel doctors of Cheshire, were ignored. A much easier road was taken. The vast stores of fertility, which had accumulated after the long rest under grass, were cashed in and converted into corn crops. The seed so obtained saved the population from starvation, but most of the resulting straw could not be used because of the shortage of labour to handle it and of insufficient cattle to convert it into humus. The grow-more-food policy was, therefore, based on the exhaustion of the soil's capital. It is a perfect example of unbalanced farming. It is therefore certain to sow the seeds of future trouble, which will be duly registered by Mother Earth in the form of malnutrition and disease of crops, livestock, and mankind.

Chapter 9
Magnoliid Families

Annonaceae

Annonaceae, the custard apple family are a family of flowering plants consisting of trees, shrubs, or rarely lianas. With about 2300 to 2500 species and more than 130 genera, it is the largest family in Magnoliales. Seven genera, *Annona*, *Anonidium*, *Rollinia*, *Uvaria*, *Melodorum*, *Asimina*, and *Stelechocarpus* produce edible fruits. Its type genus is *Annona*. The family is concentrated in the tropics, with few species found in temperate regions.

About 900 species are Neotropical, 450 are Afrotropical, and the other species Indomalayan. Compared to the species from the Neotropics, very little is known about many species from Indomalaya. Only a few attempts have been made for the phylogeny-based reclassification of the family, and those have been hampered by the Neotropic bias in the available information, with the most of the work having been done on genera and tribes.

Description

Mostly tropical, some mid-latitude, deciduous or evergreen trees and shrubs, some lianas, with aromatic bark, leaves, and flowers.

Stems, Stalks and Leaves

Bark is fibrous and aromatic. Pith septate (fine tangential bands divided by partitions) to diaphragmed (divided by thin partitions with openings in them). Branching distichous (arranged in two rows/on one plane) or spiral. Leaves are alternate, two-ranked, simple, pinnately veined, and have leaf stalks. Stipules absent.

Flowers

Flower stalks are axillary to (on the opposite side of shoot from) leaf scars on old wood and sometimes from leaves on new shoots. The flowers are usually trimerous; borne singly or in compound inflorescences; bisexual and rarely unisexual. The receptacle might become enlarged, elevated or flat. The outer whorls are inserted below the ovaries, and have valvate (overlapping) or imbricate (nonoverlapping) segments. Usually two to four persistent sepals that are distinct or connate (fused) at the base. Six petals in two unequal whorls of three with larger outer whorls and fleshier inner whorls that might share the same nectar glands, or six to fifteen petals, with impressed veins on their inner face. Ten to twenty (or many more) stamens inserted below the ovary, spirally arranged and forming a ball or flat-topped mass with short and stout filaments and linear to oblong anthers which face outward and open longitudinally. Each flower can have from one to many pistils, distinct to connate, with stigmas distinct. Marginal placentation, each pistil bearing one locule, with one to many ovules. Style short and thick, with terminal stigma.

Fruits and Seeds

Fruits are distinct (berries) or coalesce (into syncarps). Seeds are one to many per pistil; have a fleshy and usually brightly coloured cover, have ruminate endosperm (nutritive tissue surrounding the embryo) and are oily.

Systematics

Monophyly and inter-familial systematics have been well supported for Annonaceae by a combination of morphological and molecular evidence. The APG II system places Annonaceae as most closely related to the small Magnoliid family Eupomatiaceae.

Combined plastid DNA marker and morphological character analysis of *Anaxagorea* and other Annonaceae and basal angiosperm genera suggest that *Anaxagorea* is an ancestral clade, sister to the rest of the family. This may confirm the hypothesis that morphological traits shared with other

Magnoliales species (such as 2-ranked phyllotaxis, monosulcate pollen, and laminate stamens) represent ancestral characters, while derived characters observed in other genera have evolved independently multiple times. These analyses also support a biogeographic history in which Annonaceae, and perhaps other basal angiosperm clades, originated in South America during the Cretaceous, radiating due to migration events throughout Africa to Asia and Central to North America. *Guatteria*, with its approximately 265 species, is the largest genus of Annonaceae, and might be broken into three small genera based on molecular characters from multiple locations.

Tribe Saccopetaleae is another group of Annonaceae that is awaiting review as recent molecular studies suggest that this tribe is not monophyletic at all.

Uses

Food: The large, edible, pulpy fruits of some members typically called "anona" by Spanish and Portuguese speaking people of the family's Neotropical range, include species of *Annona*: custard apple (*A. reticulata*), cherimoya (*A. cherimola*), soursop/guanábana/graviola (*A. muricata*), sweetsop (*A. squamosa*), ilama (*A. diversifolia*), soncoya (*A. purpurea*), atemoya (a cross between *A. cherimola* and *A. squamosa*); and biriba (*Rollinia deliciosa*, which may require reclassification under *Annona.*).

The names of many of those fruits are sometimes used interchangeably. Recently, consumption of the neotropical annonaceous plant *Annona muricata* (soursop, graviola, guanabana) has been strongly associated as a causal agent in "atypical Parkinsonsism." The causative agent, annonancin, is present in many of the annonaceae. It is thought to be responsible for up to 70% of Parkinsonian conditions in Guadeloupe. Exposure is typically through traditional food and "natural" medicines. *Asimina triloba* (American pawpaw, prairie banana) has an Eastern U.S. distribution, and is currently under agricultural investigation as a commercial crop.

Folk Medicine

The bark, leaves, and roots of some species are used in folk medicines.

Lancewood

Lancewood is a tough, elastic, and heavy wood obtained from the from species of the family Annonaceae located in West Indies and The Guianas. It was often used especially for carriage shafts. It is brought into commerce in the form of taper poles of about 6 metres in width and from 15 to 20 cm in breadth at the butt.

The black lancewood or carisiri of The Guianas is of remarkably slender form. The Yellow lancewood tree is found in tolerable abundance throughout The Guianas, and used by the Amerinds for arrow-points, as well as for spars, beams, etc. Some bowyers use this wood for making longbows but use the term *lemonwood* or *degame* to label the wood the bow is made from.

Other

- Some species of the family, such as *Cananga odorata* (ylang-ylang) also have aromatic oil and are used for perfumes or spices.
- The strong bark is used for carrying burdens in the Amazon Rainforest and for wooden implements, such as tool handles and pegs. The wood is valued as firewood.
- Yellow and brown dyes.
- Some species are also grown as ornamental plants, especially the Indian species *Polyalthia longifolia pendula.*

Chemical Constituents

Pharmaceutical research has found antifungal, bacteriostatic, antimalarial, and especially cytostatic capability of some chemical constituents of the leaves and bark. A large number of chemical compounds, including flavonoids, alkaloids, and acetogenins, have been extracted from the seeds and many other parts of these plants. Flavonoids and alkaloids have shown antibacterial properties, and have been used for treatment of medical conditions, such as skin disease, intestinal worms, and inflammation of the eye. Potential pharmaceutical products are currently in early experimental trials in animal models and cell-lines. Acetogenins are thought to have anti-HIV and anti-cancer effects. A wide variety of products have been developed and are available for cancer treatment.

Flavonoids and alkaloids contained in the leaves and bark of several species of the family have shown insecticidal properties.

Genera

- *Afroguatteria*
- *Alphonsea*
- *Ambavia*
- *Anaxagorea*
- *Ancana*
- *Annickia*
- *Annona*
- *Anomianthus*
- *Anonidium*
- *Artabotrys* - trail-grape
- *Asimina*
- *Asteranthe*
- *Balonga*
- *Bocagea*
- *Bocageopsis*
- *Boutiquea*
- *Cananga* - Ylang-ylang
- *Cardiopetalum*
- *Cleistochlamys*
- *Cleistopholis*
- *Craibella*
- *Cremastosperma*
- *Cyathocalyx*
- *Cyathostemma*
- *Cymbopetalum*
- *Dasoclema*
- *Dasymaschalon*
- *Deeringothamnus* - false pawpaw
- *Dendrokingstonia*
- *Dennettia*

- *Desmopsis*
- *Desmos*
- *Diclinanona*
- *Dielsiothamnus*
- *Disepalum*
- *Duckeanthus*
- *Duguetia*
- *Ellipeia*
- *Ellipeiopsis*
- *Enicosanthum*
- *Ephedranthus*
- *Exellia*
- *Fissistigma*
- *Fitzalania*
- *Friesodielsia*
- *Froesiodendron*
- *Fusaea*
- *Gilbertiella*
- *Goniothalamus*
- *Greenwayodendron*
- *Guamia*
- *Guatteria* - haya minga, haya blanca
- *Guatteriella*
- *Guatteriopsis*
- *Haplostichanthus*
- *Heteropetalum*
- *Hexalobus*
- *Hornschuchia*
- *Isolona*
- *Letestudoxa*
- *Lettowianthus*
- *Malmea*
- *Marsypopetalum*

- *Meiocarpidium*
- *Meiogyne*
- *Melodorum*
- *Mezzettia*
- *Mezzettiopsis*
- *Miliusa*
- *Mischogyne*
- *Mitrella*
- *Mitrephora*
- *Mkilua*
- *Monanthotaxis*
- *Monocarpia*
- *Monocyclanthus*
- *Monodora*
- *Mosannona*
- *Neostenanthera*
- *Neo-uvaria*
- *Onychopetalum*
- *Ophrypetalum*
- *Oreomitra*
- *Orophea*
- *Oxandra* - blacklancewood, vernacular name:haya
- *Pachypodanthium*
- *Papualthia*
- *Petalolophus*
- *Phaeanthus*
- *Phoenicanthus*
- *Piptostigma*
- *Platymitra*
- *Polyalthia*
- *Polyceratocarpus*
- *Popowia*
- *Porcelia*

- *Pseudartabotrys*
- *Pseudephedranthus*
- *Pseudoxandra*
- *Pseuduvaria*
- *Pyramidanthe*
- *Raimondia*
- *Reedrollinsia*
- *Richella*
- *Rollinia* - wild sugar-apple
- *Ruizodendron*
- *Sageraea*
- *Sanrafaelia*
- *Sapranthus*
- *Schefferomitra*
- *Sphaerocoryne*
- *Stelechocarpus* - kepel
- *Stenanona*
- *Tetrameranthus*
- *Toussaintia*
- *Tridimeris*
- *Trigynaea*
- *Trivalvaria*
- *Unonopsis*
- *Uvaria*
- *Uvariastrum*
- *Uvariodendron*
- *Uvariopsis*
- *Woodiellantha*
- *Xylopia*

Toxicology

The compound annonacin contained in the seeds of some members of Annonaceae such as *Annona muricata* (soursop) is a neurotoxin and it seems to be the cause of a neurodegenerative disease. The only group of people known to be affected live on the Caribbean island of Guadeloupe and the problem presumably occurs with the consumption of plants containing annonacin. The disorder is a so-called tauopathy associated with a pathologic accumulation of tau protein in the brain. Experimental results demonstrated for the first time that the plant neurotoxin annonacin is responsible for this accumulation.

Adenodaphne

Adenodaphne is a genus of evergreen trees or shrubs belonging to the family Lauraceae, from New Caledonia. There are five species; they are large trees characteristics of Rainforest in montane laurel forest habitats in New Caledonia and restricted to this region. They belong to an ancient Gondwanian element of Laurales. They are having 12 chromosomes.

Overview

They are leafy canopy trees with erect or spreading branches. It grows to heights of up to 25 metres some species 40 m. and have stout trunks up to 1 metre in diametre. The thick, leathery leaves are dark green. The leaves are glossy laurel type.

The forests are made up of laurel-leaved evergreen hardwood trees, harbouring a rich biota of understorey plants, invertebrates, birds and bats. *Adenodaphne* require continuously moist soil, and do not tolerate drought.

Trees or shrubs, dioecious. The ecological requirements of the genus, are those of the laurel forest and like most of their counterparts laurifolia in the world, they are vigorous species with a great ability to populate the habitat that is conducive. Adenodaphne genus responde to characteristic formations of laurel forest.

The natural habitat is tropical montane laurel forest of New Caledonia which are cloud-covered for much of the year. These species are found in forests that face threats of destruction by human deforestation. It could caused the extinction of the genus across the restricted region area, resulting in the present distribution. They are in danger of extinction due to over exploitation as medicinal plants or timber extraction and also for loss of habitat.

Characteristics

The genus includes species of trees, and shrubs, with evergreen foliage and inconspicuous flowers. Adenodaphne are dioecious and have mostly smooth, glossy, lauroid type leaves. They are evergreen tree with some species growing to 25 m tall.

The inflorescences are consisting in pseudo-umbels (a flat-topped or rounded flower cluster) that are arranged in a racime, sometimes condensed, or a short-shoot or rarely sessile. Each pseudo-umbel with an involucre of decussate, crossed in the form of an X, usually persistent bracts. Leaves glabrous or pubescent, domatia absent.

Inflorescences axillary or solitary seudoumbelas along very short sharp branches, appearing racemose, covered before anthesis by an involucre of bracts decussate.

The flower is from greenish, yellow to white. Male and female flower on different plants. The pollination is done by bees and other insects. The flowers are irregular. The flowers are unisexuals, male flowers with 5 to 20 fertile stamens. The female with a ovary globose and a floral tube. The flowers could be without petals to nine petals by species. The petals when present are equal or unequal, often caducous during anthesis. In the stamens, several inner ones with glands.

Filaments usually longer than the anthers. Anthers are having four locules. The pollen sacs arranged mostly in two pairs above each other, all introrse. Lower pollen sacs are latrorse. The staminodes usually are absent. The tepals usually are deciduous. The Pistillo or gynoecium well developed to absent. Fruits on a light or markedly thickened pedicel, supported by a shallow or deep dome, simple margin. The flower have a flat to deeply cup-shaped receptacle. The fruit is a drupe of variable shape and size. The most striking are its fruits, The fruit is an ellipsoid to ovoid drupe or berry, and the seed is a single kernel. Plum-like to olive-like drupe settled on a discoid small dome. With shape rounded or ovoid, brown to black, rarely green, purple, reddish, orange to pale yellow. The fruits are a very important food source for birds and other wildlife. The seed dispersion is the result of scattering by columbiform birds mostly but also bats and endemics birds like kagu, parrots or rallidaes.

Most seeds pass through the bird's digestive system intact. Seed dispersal via ingestion by vertebrate animals, mostly birds and mammals, is the dispersal mechanism for most tree species.

Aristolochiaceae

The Aristolochiaceae, or the Birthwort family, are a family of flowering plants with 7 genera and about 400 species belonging to the order Piperales. The type genus is *Aristolochia* L.

Description

They are mostly perennial, herbaceous plants, shrubs, woody vines or even lianas. The membranous, cordate simple leaves are spread out, growing alternately along the stem on leaf stalks. The margins are commonly entire. There are no stipules. The bizarre flowers are large to medium-sized, growing in the leaf axils. They are bilaterally or radially symmetrical.

Classification

The Aristolochiaceae are magnoliids, a basal group of angiosperms which are not part of the large categories of monocots or eudicots. Some newer classification schemes, such as the Update of the Angiosperm Phylogeny Group, place the family Aristolochiaceae in the order Piperales, but it is still quite common, though superseded, for the Aristolochiaceae to be assigned, sometimes with some other families, their own order (Aristolochiales).

Phytochemistry

Many members of *Aristolochia* and some of *Asarum* contain the toxin aristolochic acid, which discourages herbivores and is known to be carcinogenic in rats. *Aristolochia* itself is carcinogenic to humans.

Ecology

Pipevine swallowtail butterflies lay their eggs on pipevine (*Aristolochia* species), and the larvae dine on the plant but are not affected by the toxin, which then offers the adult butterfly protection against predators.

Atherospermataceae

The Atherospermataceae, commonly known as the southern sassafrases, are a family of broadleaf evergreen trees and shrubs.

The family includes 14 species in seven genera. The atherosperms are of Gondwanan origin, and are today mostly distributed in the southern hemisphere, with two species native to southern Chile and twelve species native to Australasia. Wood is commercially harvested from rainforest species of this family, and is used both in construction and in fine cabinet making.

Ecology

These trees and shrubs are characteristic of the lower strata of the tropical rainforest, except *Dryadodaphne* species which belong to the rainforest high canopy. The glands at the base of the stamens secrete nectar in *Laurelia novae-zelandiae*, which accumulates at the base of the flower and attracts bees, coleoptera and Bombyliidae. The seed, in the form of a feathery achene is dispersed by wind (anemochory).

The wood of *Laurelia* has local interest for construction, particularly the Chilean *Laurelia sempervirens*, despite its lack of resistance to moisture. Essential oils extracted from the leaves and bark of species of *Doryphora* have application in perfumery and pharmaceuticals.

Evolution

The *Atherospermataceae* have in the past been treated as a subfamily (*Atherospermatoideae*) of the *Monimiaceae*. Recent reassessment of both morphological and molecular characters, however, show them to be more clearly related to the *Gomortegaceae* and *Siparunaceae*. The APW (Angiosperm Phylogeny Website) considers them to be a family of their own (as the *Atherospermataceae*), and together with the *Gomortegaceae* and *Siparunaceae* they form a distinct branch of the *Laurales*.

A clade made up of *Doryphora* and *Daphnandra*, from the Australian states of Queensland and New South Wales, is considered sister to the rest of the family. The remaining genera (*Atherosperma*, *Dryadodaphne*, *Laurelia*, *Laureliopsis* and *Nemuaron*) are considered by some to be more advanced evolutionarily.

The atherosperm fossil record, which goes back to the Upper Cretaceous, includes pollen, wood, and leaf fossils from Europe, Africa, South America, Antarctica, New Zealand, and

Tasmania. The family is recognised in Antarctic fossils from the Cretaceous to the lower Tertiary (about 88 mya). Fossil pollen of *Laurelia* has been attributed to the middle Oligocene of New Zealand, as well as the Eocene-Oligocene and Early Miocene of Argentina and Seymour Island.

Calibration of rbcL substitution rates within the fossils suggests an initial diversification of the family at 100-140 mya, probably in West Gondwana, early entry into Antarctica, and long-distance dispersal to New Zealand and New Caledonia at 50-30 mya by the ancestors of *Laurelia novae-zelandiae* and *Nemuaron*.

Species

- *Atherosperma moschatum* (Tasmania, Victoria and New South Wales)
- *Atherosperma moschatum subsp. integrifolium* (New South Wales)
- *Daphnandra apatela* (Queensland and New South Wales)
- *Daphnandra micrantha* (New South Wales)
- *Daphnandra johnsonii* (New South Wales)
- *Daphnandra repandula* (Queensland)
- *Daphnandra tenuipes* (New South Wales)
- *Doryphora aromatica* (northeast Queensland)
- *Doryphora sassafras* (Queensland and New South Wales)
- *Dryadodaphne crassa* (New Guinea)
- *Dryadodaphne novoguineensis* (New Guinea)
- *Dryadodaphne trachyphloia* (Queensland)
- *Laurelia novae-zelandiae* (New Zealand)
- *Laurelia sempervirens* (southern Chile)
- *Laureliopsis philippiana* (southern Chile and Argentina)
- *Nemuaron vieillardii* (New Caledonia)

Calycanthaceae

The Calycanthaceae (sweetshrub or spicebush family) is a small family of flowering plants included in the order Laurales.

The family contains three genera and only 10 species, restricted to warm temperate and tropical regions:

- *Calycanthus* (2 species; western and southeastern North America)
- *Chimonanthus* (6 species; eastern Asia)
- *Idiospermum* (1 species; Queensland, Australia)
- *Sinocalycanthus* (1 species; eastern Asia) - morphological and molecular data indicate that this is a member of Calycanthus

They are aromatic deciduous shrubs growing to 2–4 m tall, except for *Idiospermum*, which is a large evergreen tree. The flowers are white to red, with spirally arranged tepals.

DNA-based phylogenies indicate that the Northern Hemisphere Calycanthus and Chimonanthus diverged from each other in the mid-Miocene, while the Australian Idiospermum had already diverged by the Upper Cretaceous and likely represents a remnant of a former Gondwanan distribution of Calycanthaceae that included South America, as indicated by the occurrence of Cretaceous Calycanthaceae fossils in Brazil.

Degeneriaceae

Degeneriaceae is the botanical name for a family of flowering plants. Such a family has been recognised by more than a few taxonomists, at least over the past few decades. The APG II system, of 2003 (unchanged from the APG system, of 1998), also recognises this family, and assigns it to the order Magnoliales in the clade magnoliids. The family consists of one genus only, *Degeneria*, of one or two species, trees, found on Fiji.

Hernandiaceae

Hernandiaceae is the botanical name for a family of flowering plants. They are a family of angiosperms in order Laurales. Consisting of 5 genera with about 59 species, distributed by the world's tropical areas, some of them widely distributed in coastal areas but they are occurring from sea level to over 2000 metres.

The family is closely related to Lauraceae, and many species inhabit laurel forest habitat and they are having laurel-like (lauroid) leaves. Based on morphology, chromosome numbers, geographical distribution and phylogenetic analyzes, the family is clearly divided into 2 groups that have been given the rank of subfamilies Gyrocarpoideae and Hernandioideae.

Overview

The Hernandaceae are important components of tropical forests ranging from low-lying to montane. Because of the special lack of worldwide knowledge about the family in general, little is known about their diversity yet. The knowledge of this family to national level, is that to be expected in countries with limited economic means with the vast majority of species indeterminate or even poorly determined. Recently described new species come from collections made in countries with limited economic means.

Therefore an increase in the study of family, at national level, is of utmost importance for the progress of the systematics of the family in general. Trees of the Hernandiaceae family predominate in the world's laurel forests and cloud forest, which occur in tropical, subtropical and mild temperate regions of the northern and southern hemispheres, highlighting the African, Indic and Pacific ocean islands, New Caledonia, Madagascar, and central Chile. There are three main economical uses for this family. A high content of essential oils are found in many species that are important for spices and perfumes. The hard wood of many species is a source for timber around the world. A great number of species are in danger of extinction due to over exploitation as medicinal plants or timber extraction and also for loss of habitat.

Description

Trees, or shrubs, or lianas; bearing essential oils. Self-supporting, or climbing. Leaves alternate; aromatic; simple, or compound; peltate (often, in Hernandia), or not peltate; if compound, palmate. Lamina dissected, or entire; when dissected, palmatifid; palmately veined; cross-venulate. Leaves estipulate. Cork cambium present; initially superficial. Nodes unilacunar. Primary vascular tissue in a cylinder, without separate bundles. Cortical bundles absent. Medullary bundles absent. Internal phloem absent. Secondary thickening developing from a conventional cambial ring. Included phloem absent. Xylem with libriform fibres. Vessel end-walls simple. Vessels without vestured pits. Wood parenchyma paratracheal. Sieve-tube plastids P-type; type I.

Ecology

The Hernandiaceae is a very important family as their species are a large number of individuals inhabiting the ecosystems with monoecious, rarely dioecious, deciduous or evergreen trees, shrubs and perennial climbing plants. The mode of dispersion is variable among the species of the genera. Most species of genus Hernandia have red domes suggesting zoochory, while Hernandia guianensis is hydrochorian in freshwater, and Hernandia nymphaeifolia and Gyrocarpus americanus are hydrochorian in seawater.

There are fruits that are eaten by birds, passing through the digestive tract, or are regurgitated germinating after easily. Zoocorian seeds are mostly in small fruits, desired by birds, containing many times a sticky substance coating the seed. Expelled, the seed sticks to the branch of trees and even dead bodies, or just in soil, germinating later. There are also fruits that open very violently expelling the seeds away. Other are nuts o non-fleshy bodies with small size (achene) these are provided with hooks or filaments that stick to the fur of animals or shaped to float in water or facilitate transport by wind. They are distributed in the lower areas of the tropics, especially in rain forests, Cloud forest and Laurissilva although some species exist even in subtropical or arid areas and they are occurring from sea level to over 2000 metres. Intra distributions in Africa and America for example from Gyrocarpus hababensis and Gyrocarpus americanus, relict character, appear to be due to marine transgressions in the past.

The family originated in the coastal laurel forests of Gondwana, which is the reason for its pattern of distribution. The Hernandaceae inhabit montane tropical forests, some species reaching 4000 m above sea level, but most species are more frequent in low-altitude rainforests. Some deciduous species have adapted to demanding conditions in semi-arid climates but then they tend to depend on favourable edaphic conditions, for example, perennial aquifers, periodic groundwater flows, or periodically flooded forests in sand containing hardly any nutrients.

Flowers and Fruit

The plants can be hermaphrodite, or monoecious. The flowers are aggregated in inflorescences, in cymes, regular, cyclic or tetracyclic. They present a perianth with distinct calyx and corolla. The corolla is not fleshy. Androecial members free of the perianth; free of one another; 1 whorled. Androecium exclusively of fertile stamens, or including staminodes (these in the form of 1–2 nectariferous glands outside the stamens, cf. Lauraceae). Staminodes external to the fertile stamens (3–5); oppositisepalous (alternating with the inner perianth whorl). Anthers dehiscing by longitudinal valves. Anther wall of the 'dicot' type. Pollen grains nonaperturate; 2-celled. Gynoecium 1 carpelled. Carpels reduced in number relative to the perianth. The pistil 1 celled. Gynoecium monomerous; of one carpel; inferior. Carpel stylate; apically stigmatic; 1 ovuled. Placentation apical. Ovules pendulous; anatropous; bitegmic; crassinucellate. Endosperm formation cellular.

The fruit in some species is non-fleshy and presents the following characteristics: fruiting carpel indehiscent, winged or enclosed in an inflated envelope derived from connate bracteoles, with only 1 seed non-endospermic, 2 cotyledons (fleshy, oily, smooth or ruminate). The embryo is straight.

Taxonomy

The family has been recognised by most taxonomists. The APG II system, of 2003 (unchanged from the APG system, of 1998), also recognises this family, and assigns it to the order Laurales in the clade magnoliids. As circumscribed by APG the family includes those plants that sometimes have been treated as forming the family Gyrocarpaceae. The family is usually described as counting four genera, totalling several dozen species.

Phytochemistry

Not cyanogenic. Alkaloids present (Efatine, (+)-O,O-dimethyl corytuberine, hernandaline, hernandine, hernanymphine and herveline A). Iridoids not detected. Arthroquinones detected (*Illigera*); polyacetate derived. Proanthocyanidins absent. Lignans present (Deoxypodophyllic acid, (+)-hernangerine, hernolactone, epimagnolin, epieudesmine, epiaschantin). Flavonols present: kaempferol and quercetin. Ellagic acid absent (*Hernandia*). Saponins/sapogenins absent. Aluminium accumulation not found. Sugars transported as sucrose (*Hernandia*).

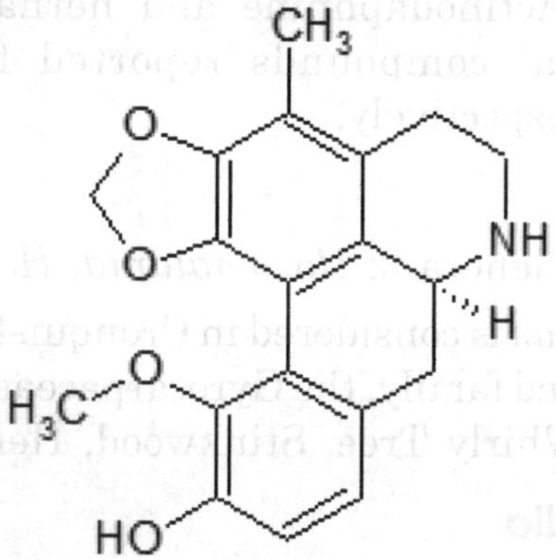

Figure: *Hernandine, aporphine alkaloid from Hernandia*

Uses

The bark, seeds and young leaves are purgative. The root is chewed as a remedy against eating poisonous crabs and fishes. The juice of the bark and leaves has depilatory properties. The available information on the plant parts used by Samoan healers, their modes of preparation and application have been listed in Table-1. The leaves of the plant are used as hypertensive antitumor prophylaxis of arteriosclerosis piscicide.

The Stem bark, root bark, twigs, stalks and seeds of the plant are used as hypotensive, non-cholinergic, relaxant smooth musculature, vasopressive, sympathicomimetic for inflammation and irritation of the upper respiratory system and the gastrointestinal tract. Also used as mild laxative, antidiarrheal, cytotoxic, cytostatic, antiviral, externally on boils, ulcers, sores, inflammations. The pericarp of the fruits also relaxes the smooth musculature, produces a cataleptic action, anti-convulsive, in convulsive disorders, vasopressive, reduces intestinal movement, hypotensive, non-cholinergic, dilation of capillary blood vessels (nontoxic low doses) respiration is accelerated, uterus-stimulating (large doses) respiration is slowed, uterus- inhibitor, stimulation of the secretion of tears and saliva, might cause emesis and moderate tachycardia, hypnotic, sedative, narcotic, adrenolytic, psychomimetic, induces hyperglycemia.

Among the chemical compounds isolated from the family Hernandiaceae, the alkaloid corytuberine is the oldest known compound. Later on, its derivative O,O-Dimethylcorytuberine was reported from several *Hernandia* species, including *H. nymphaeifolia*. Actinodaphnine and hernandion, were the earliest chemical compounds reported from the family hernandiaceae respectively. .

Species

Species 54. Genera 3; *Hazomalania*, *Hernandia*, *Illigera*.

- *Gyrocarpus* is considered in Cronquist System to belong a separated family, the Gyrocarpaceae. (Propellerbaum, Whirly Whirly Tree, Stinkwood, Helicopter Tree)

Genus Hernandia

- *Hernandia ovigera* (syn. *H. peltata*) is the most abundant species in terms of the sheer number of compounds isolated and subsequent biological activities reported in the literature.
- *Hernandia albiflora*
- *Hernandia beninensis*
- *Hernandia bivalvis*
- *Hernandia catalpifolia*
- *Hernandia cordigera*
- *Hernandia cubensis*
- *Hernandia didymantha*
- *Hernandia drakeana*
- *Hernandia guianensis*
- *Hernandia hammelii*
- *Hernandia jamaicensis*
- *Hernandia kunstleri*
- *Hernandia leucomphala*
- *Hernandia lychnifera*
- *Hernandia mascarenensis*
- *Hernandia moerenhoutiana* Guill. (indigenous to Samoa Islands, also occurs westward to Melanesia.)

- *Hernandia nukuhivensis*
- *Hernandia nymphaeifolia*
- *Hernandia obovata*
- *Hernandia olivacea*
- *Hernandia papuana*
- *Hernandia peltata*
- *Hernandia rostrata*
- *Hernandia samoensis*
- *Hernandia sonora*
- *Hernandia stenura*
- *Hernandia stokesii*
- *Hernandia tahitensis*
- *Hernandia temarii*
- *Hernandia voyroni*
- *Hernandia wendtii*

Genus Illigera

- *Illigera appendiculata*
- *Illigera aromatica*
- *Illigera brevistaminata*
- *Illigera cardiophylla*
- *Illigera celebica*
- *Illigera cordata*
- *Illigera coryzadenia*
- *Illigera cucullata*
- *Illigera dasyphylla*
- *Illigera diptera*
- *Illigera dubia*
- *Illigera dunniana*
- *Illigera elegans*
- *Illigera elliptifolia*
- *Illigera fordii*
- *Illigera gammiei*

- *Illigera glabra*
- *Illigera glandulosa*
- *Illigera grandiflora*
- *Illigera henryi*
- *Illigera khasiana*
- *Illigera kurzii*
- *Illigera lucida*
- *Illigera luzonensis*
- *Illigera madagascariensis*
- *Illigera megaptera*
- *Illigera meyeniana*
- *Illigera mollissima*
- *Illigera nervosa*
- *Illigera novoguineensis*
- *Illigera obtusa*
- *Illigera orbiculata*
- *Illigera ovatifolia*
- *Illigera parviflora*
- *Illigera pentaphylla*
- *Illigera petelotii*
- *Illigera pierrei*
- *Illigera platyandra*
- *Illigera pseudoparviflora*
- *Illigera pubescens*
- *Illigera pulchra*
- *Illigera reticulata*
- *Illigera rhodantha*
- *Illigera rhodanthe*
- *Illigera ternata*
- *Illigera thorelii*
- *Illigera trifoliata*
- *Illigera vespertilio*

- *Illigera villosa*
- *Illigera yaoshanensis*

Genus Hazomalania

- *Hazomalania voyronii*

Himantandraceae

Himantandraceae is the botanical name for a family of flowering plants. Such a family has been recognised by most taxonomists. The APG II system, of 2003 (unchanged from the APG system, of 1998), also recognises this family, and assigns it to the order Magnoliales in the clade magnoliids. The family consists of one genus only, *Galbulimima*, of probably two species, trees and shrubs, found in tropical areas in Southeast Asia and Australia.

Hydnoraceae

Hydnoraceae is a family of parasitic flowering plants in the order Piperales. It contains two genera, *Hydnora* and *Prosopanche* and some seven species. *Prosopanche* contains two species from Central and South America and *Hydnora* contains five species from arid areas of Africa, Saudi Arabia, and Madagascar.

Description

The most striking aspect of the Hydnoraceae is probably the complete absence of leaves (not even in modified forms such as scales). Some species are mildly thermogenic (capable of producing heat), presumably as a means to dispersing their scent.

Ecology

The plants are pollinated by insects such as dermestid beetles or carrion flies, attracted by the fetid odor of the flowers. In *Hydnora africana* there are *bait bodies* with a strong smell, whereas in *Hydnora johannis* the scent comes from a region at the tip of the perianth called a *cucullus*. The flowers may be above ground or underground. The fruits have edible, fragrant pulp, which attracts animals such as porcupines, monkeys, jackals, rhinoceros, and armadillos, as well as humans. The host plants, in the case of *Hydnora*, generally are in the family Euphorbiaceae and the genus *Acacia*. Those for *Prosopanche* include various species of *Prosopis* and other legumes.

Biochemistry: The plants contain high levels of tannins.

Classification

Like many parasitic plants, the affinities with non-parasitic plants are not obvious, and 19th and 20th century botanists proposed a variety of placements for the family. Molecular data places them in the Piperales, more closely related to Aristolochiaceae than to Piperaceae or Saururaceae.

Lauraceae

The Lauraceae are the laurel family, that includes the "True Laurel" and its closest relatives. The family comprises over 3000 species of flowering plants in over 50 genera worldwide. They occur mainly in warm temperate and tropical regions, especially Southeast Asia and South America. Most are aromatic evergreen trees or shrubs, but one or two genera such as *Sassafras* are deciduous, and *Cassytha* is a genus of parasitic vines.

Overview

The family has a worldwide distribution in tropical and warm climates. The Lauraceae are important components of tropical forests ranging from low-lying to montane. In several forested regions Lauraceae are among the top five families in terms of the number of species present.

The Lauraceae give their name to habitats known as Laurel forests, which have many trees that superficially resemble the Lauraceae, though they may belong to other plant families such as Magnoliaceae or Myrtaceae. Laurel forests of various types occur on most continents and on many major islands.

Although the taxonomy of the Lauraceae is still not settled, conservative estimates suggest some 52 genera worldwide, including 3000 to 3500 species. Compared to other plant families, the taxonomy of Lauraceae still is poorly understood. This is partly due to its great diversity, the difficult of identifying the species, and partly because of inadequate investment in taxonomic work.

Recent monographs on small and medium-sized genera of Lauraceae (up to about 100 species), have revealed many new species. Similar increases in the numbers of species recognised in other larger genera are to be expected.

Description

Most of the Lauraceae are evergreen trees in habit. Exceptions include some two dozen species of *Cassytha*, all of which are obligately parasitic vines.

The fruit of Lauraceae is a drupe, a one-seeded fleshy fruit with a hard layer, the endocarp, surrounding the seed. However, the endocarp is very thin, so the fruit resembles a one-seeded berry. The fruit in some species (particularly in the genera *Ocotea* and *Oreodaphne*) is partly immersed or covered in a cup-shaped or deep thick cupule, which is formed from the tube of the calyx where the peduncle joins the fruit; this gives the fruit an appearance similar to an acorn. In some *Lindera* species the fruit have a hypocarpium at the base of the fruit.

Distribution and Uses

Because the family is so ancient and was so widely distributed on the Gondwana supercontinent, modern species commonly occur in relict populations isolated by geographical barriers, for instance on islands or tropical mountains. Relict forests retain endemic fauna and flora in communities of great value in inferring the palaeontological succession and climate change that followed the breakups of the supercontinents.

- Many Lauraceae contain high concentrations of essential oils, some of which are valued for spices and perfumes. Within the plants themselves most such substances are components of irritant or toxic sap or tissues that repel or poison many herbivorous or parasitic organisms. Some of the essential oils are valued as fragrances, such as in the traditional laurel wreath of classical antiquity, or in cabinet making, where the fragrant woods are prized for making insect-repellant furniture chests. Some are valued in cooking, for example, bay leaves are a popular ingredient in European, American and Asian cuisines.
- Avocados are important oil-rich fruit that are cultivated in warm climates around the world.
- Many species are exploited for timber.
- Some species are valued as sources of medicinal material.

The following genera include some of the best known species of particular commercial value:

- *Cinnamomum*: Cinnamon, Cassia and Camphor Laurel
- *Laurus*: Bay Laurel
- *Persea*: Avocado

Loss of habitat and over-exploitation for such products has put many species in danger of extinction as a result of overcutting, extensive illegal logging and habitat conversion.

Conversely, some species, though commercially valuable in some countries, are regarded as aggressive invaders in other regions. For example, *Cinnamomum camphora*, though a valued ornamental and medicinal plant, is so invasive as to have been declared a weed in subtropical forested areas of South Africa.

Ecology

Lauraceae flowers are protogynous, often with a complex flowering system to prevent inbreeding. The fruits are an important food source for birds, on which some Palaeognathae are highly dependent. Other birds that rely heavily on the fruit for their diet include members of the families Cotingidae, Columbidae, Trogonidae, Turdidae and Rhamphastidae, amongst others. Birds that are specialised frugivores tend to eat the whole fruit and regurgitate seeds intact, thereby releasing the seeds in favourable situations for germination (ornithochory). Some other birds that swallow the fruit pass the seed intact through the gut. Seed dispersal of various species in the family is also carried out by monkeys, arboreal rodents, porcupines, opossums and fishes. Hydrochory occurs in *Caryodaphnopsis*.

The leaves of some species in the Lauraceae have domatia in the axils of their veins. The domatia are home to certain mites. Other Lauraceous species, members of the genus *Pleurothyrium* in particular, have a symbiotic relationship with ants that protect and defend the tree. Some *Ocotea* species are also used as nesting sites by ants, which may live in leaf pockets or in hollowed-out stems.

Defence mechanisms that occur among members of the Lauraceae include irritant or toxic sap or tissues that repel or poison many herbivorous organisms.

Trees of the family predominate in the world's laurel forests and cloud forests, which occur in tropical to mild temperate regions of both northern and southern hemispheres. Other members of the family however, occur pantropically in general lowland and Afromontane forest, and in Africa for example there are species endemic to countries such as Cameroon, Sudan, Tanzania, Uganda and Congo. Several relict species in the Lauraceae occur in temperate areas of both hemispheres. Many botanical species in other families have similar foliage to the Lauraceae due to convergent evolution, and forests of such plants are called laurel forest.

These plants are adapted to high rainfall and humidity, and have leaves with a generous layer of wax, making them glossy in appearance, and a narrow, pointed-oval shape with a 'drip tip', which permits the leaves to shed water despite the humidity, allowing transpiration to continue. Scientific names similar to *Daphne* (e.g., *Daphnidium*, *Daphniphyllum*) or "laurel" (e.g.,*Laureliopsis*, *Skimmia laureola*) indicate other plant families that resemble Lauraceae. The dispersal of seeds in many laurel forest species is due to birds that swallow them, so the fruit and berries are often similar to attract birds.

Some Lauraceae species have adapted to demanding conditions in semi-arid climates but they tend to depend on favourable edaphic conditions, for example, perennial aquifers, periodic groundwater flows, or periodically flooded forests in sand that contains hardly any nutrients. Various species have adapted to swampy conditions by growing pneumatophores, roots that grow upward, that project above the levels of periodic floods that drown competing plants which lack such adaptations.

Paleobotanists have suggested that the family originated some 174±32 Ma, while others do not believe that they are older than the mid-Cretaceous. Fossil flowers attributed to this family occur in Cenomanian clays (mid-Cretaceous, 90-98 Ma ago) of the Eastern United States (*Mauldinia mirabilis*). Fossils of Lauraceae are common in the Tertiary strata of Europe and North America, but they virtually disappeared from central Europe in the Late Miocene. Because of its unusual fragility, the pollen of Lauraceae does not keep well and has been found only in relatively recent strata.

The patterns of speciation in the Lauraceae indicate that since the onset of aridification on the continents 15 million years ago, rainforests were fragmented into the current archipelagos of rainforests around the planet. This led to increases in speciation with the majority of our modern species being the products of parapatric speciation. Many of the resulting species have a very disjunct distribution with sister taxa occurring in other centres of *rainforest* and *cloud forest* diversity. This supports the idea that the group is of Gondwanan origin.

Deciduous Lauraceae lose all of their leaves for part of the year depending on variations in rainfall. The leaf loss coincides with the dry season in tropical, subtropical and arid regions.

Classification

Classification within the Lauraceae is not fully resolved. Multiple classification schemes based on a variety of morphological and anatomical characteristics have been proposed, but none are fully accepted. According to Judd et al. (2007), the suprageneric classification proposed by van der Werff and Richter (1996) is currently the authority. However, due to an array of molecular and embryological evidence that disagrees with the groupings, it is not fully accepted by the scientific community. Their classification is based on inflorescence structure and wood and bark anatomy. It divides Lauraceae into two subfamilies, Cassythoideae and Lauroideae. The Cassythoideae comprises a single genus, *Cassytha*, and is defined by its herbaceous, parasitic habit. The Lauroideae is then divided into three tribes: *Laureae*, *Perseeae*, and *Cryptocaryeae*.

The subfamily Cassythoideae is not fully supported. Backing has come from *matK* sequences of chloroplast genes while a questionable placement of *Cassytha* has been concluded from analysis of intergenetic spacers of chloroplast and nuclear genomes. Embryological studies also appear contradictory. One study by Heo et al. (1998) supports the subfamily. It found that *Cassytha* develops an *ab initio* cellular type endosperm and rest of the family (with one exception) develops a nuclear type endosperm. Kimoto et al. (2006) suggests that *Cassytha* should be placed in the Cryptocaryeae tribe because it shares a glandular anther tapetum and an embryo sac protruding from the nucellus with other members of the Cryptocaryeae.

The Laureae and Perseeae tribes are not well supported by any molecular or embryological studies. Sequences of the *matK* chloroplast gene as well as sequences of chloroplast and nuclear genomes reveal close relationships between the two tribes. Embryological evidence does not support a clear division between the two tribes either. Genera such as *Caryodaphnopsis* and *Aspidostemon* that share embryological characteristics with one tribe and wood and bark characteristics or inflorescence characteristics with another tribe blur the division of these groups. All available evidence, except for inflorescence morphology and wood and bark anatomy, fails to support separate Laureae and Perseeae tribes.

The Cryptocaryeae tribe is partially supported by molecular and embryological studies. Chloroplast and nuclear genomes supports a tribal grouping that contains all the genera circumscribed by van der Weff and Richter (1996) as well as three additional genera. Partial support for the tribe is also attained from the *matK* sequences of chloroplast genes as well as embryology.

Challenges in Lauraceae Classification

The knowledge of the species comprising the Lauraceae is incomplete. As of 1991, approximately 25-30% of neotropical Lauraceae species had not been described. As of 2001, embryological studies had only been completed on individuals from 26 genera yielding a 38.9% level of knowledge, in terms of embryology, for this family. Additionally, the huge amount of variation within the family poses a major challenge for developing a reliable classification.

Phytochemistry

The adaptation of Lauraceae to new environments has followed a long evolutionary journey which has led to many specialisations, including defensive or deterrent systems against other organisms.

Phytochemicals in the Lauraceae are numerous and diverse. Benzylisoquinoline alkaloids include aporphines, oxoaporphines, as well as derivatives of morphinans. Essential oils include terpenoids, benzyl benzoates, allylphenols and propenylphenols. Lignans and neolignans are present, along with *S*-methyl-5-O-

flavonoids, proanthocyanidins, cinnamoylamides, phenylpyrroles, styryl pyrones, polyketides (acetogenins), furanosesquiterpenes, and germacranolidous, heliangolidous, eudesmanolidous and guaianolidous sesquiterpene lactones.

Genera

Recent taxonomic revisions of the family include the following genera:

- *Actinodaphne*
- *Aiouea*
- *Alseodaphne*
- *Aniba*
- *Apollonias*
- *Aspidostemon*
- *Beilschmiedia*
- *Caryodaphnopsis*
- *Cassytha*
- *Chlorocardium*
- *Cinnadenia*
- *Cinnamomum*
- *Cryptocarya*
- *Dehaasia*
- *Dicypellium*
- *Dodecadenia*
- *Endiandra*
- *Endlicheria*
- *Eusideroxylon*
- *Gamanthera*
- *Hexapora*
- *Hufelandia*
- *Hypodaphnis*
- *Iteadaphne*
- *Kubitzkia*
- *Laurus*

- *Licaria*
- *Lindera*
- *Litsea*
- *Machilus*
- *Malapoenna*
- *Mezilaurus*
- *Misanteca*
- *Mocinnodaphne*
- *Mutisiopersea*
- *Nectandra*
- *Neocinnamomum*
- *Neolitsea*
- *Notaphoebe*
- *Nothaphoebe*
- *Ocotea*
- *Oreodaphne*
- *Parasassafras*
- *Parthenoxylon*
- *Paraia*
- *Persea*
- *Phoebe*
- *Phyllostemonodaphne*
- *Pleurothyrium*
- *Polyadenia*
- *Potameia*
- *Potoxylon*
- *Povedadaphne*
- *Ravensara*
- *Rhodostemonodaphne*
- *Sassafras*
- *Schauera*
- *Sextonia*
- *Sinopora*

- *Sinosassafras*
- *Syndiclis*
- *Tetranthera*
- *Tylostemon*
- *Umbellularia*
- *Urbanodendron*
- *Williamodendron*

The following genera that have traditionally been considered separate within Lauraceae, have not yet been recently re-evaluated:

- *Adenodaphne*
- *Temmodaphne*

Chapter 10

Monocot Families

Burmanniaceae

Burmanniaceae is a botanical name of a family of flowering plants, consisting of about a hundred species of herbaceous plants in roughly a dozen genera. Often they are quite remarkable plants, more often red than green, without much leaf area and not growing very big in any way. The reason for this appearance is that the members of the family are all mycoheterotrophic species. They are mainly distributed in the southern hemisphere and grow in shady and tropical environments. Many of the plants in the family are threatened species. The APG II system, of 2003, places the family in the order Dioscoreales, in the clade monocots. The circumscription of the family in APG II is wider than in the APG system, of 1998, and includes the plants that belonged to the family Thismiaceae in APG.

However, the older classification better reflects the evolutionary relationships between the genera. The clades are as follows:

Burmanniaceae *sensu stricto*

- *Apteria*
- *Burmannia*
- *Campylosiphon*
- *Cymbocarpa*
- *Dictyostega*
- *Gymnosiphon*
- *Hexapterella*

- *Marthella*
- *Mierisella*

Afrothismia clade

- *Afrothismia*

Tribe Thismieae

- *Haplothismia*
- *Oxygyne*
- *Thismia*
- *Tiputinia*

According to mollecular analyses the myco-heterotrophic type of life that these species lead evolved six (or even more) times independently in the three clades that are part of Burmanniaceae. *Afrothismia* and tribe Thismieae represent two of these shifts to myco-heterotrophy from autotrophy while Burmanniaceae *sensu stricto* are the clade where the other four took place.

The family appears in Late Cretaceous but the further diversification and shifts to the typical habit occurred later in the same period and continued after the K-T boundary in Paleogene.

Cyclanthaceae: Cyclanthaceae is a family of flowering plants.

Classification

The APG system (1998) and the APG II system (2003) assign it to the order Pandanales in the clade monocots. The family occurs in the neotropics and consists of 12 genera.

Genera

- *Asplundia* Harling
- *Carludovica* Ruiz & Pav.
- *Chorigyne* R.Erikss.
- *Cyclanthus* Poit.
- *Dianthoveus* Hammel & Wilder
- *Dicranopygium* Harling
- *Evodianthus* Oerst.
- *Ludovia* Brongn.
- *Schultesiophytum* Harling

- *Sphaeradenia* Harling
- *Stelestylis* Drude
- *Thoracocarpus* Harling

Cultivation and Uses

The family is probably best known for *Carludovica palmata*, the young leaves of which are made into Panama hats.

An unidentified species belonging to this family (possibly a *Carludovica* species) has been marketed as a houseplant in the United States under the name "Jungle Drum". *Carludovica divergens* is added to some versions of the hallucinogenic drink Ayahuasca.

Dioscoreaceae

Dioscoreaceae is a family of monocotyledonous flowering plants, with about 750 species in eight or nine genera.

The best-known member of the family is the Yam (*Dioscorea*). The APG system (1998) and APG II system (2003) both place it in the order Dioscoreales, in the clade monocots.

However, the circumscription changed in the APG II system, with the 2003 system expanded to include the plants that in the 1998 system were treated in the families Taccaceae and Trichopodaceae.

Genera

Dioscoreaceae (*sensu stricto*)

- *Borderea*
- *Dioscorea*
- *Epipetrum*
- *Rajania*
- *Stenomeris*
- *Tamus* (often included in *Dioscorea*)

(Taccaceae)

- *Tacca*

(Trichopodaceae)

- *Avetra*
- *Trichopus*

Nartheciaceae

Nartheciaceae is a botanical name of a family of flowering plants. Such a family has only rarely been recognised by systems of plant taxonomy.

The APG III system does recognise such a family and places it in the order Dioscoreales, in the clade monocots. As circumscribed by APG III it includes a few dozen species of herbaceous plants in four or five genera, including:

- *Aletris*
- *Lophiola*
- *Metanarthecium*
- *Narthecium*
- *Nietneria*

Pandanaceae

The Pandanaceae are a family of flowering plants native to the tropics of the Old World. Such a family has been widely recognised by taxonomists. Pandanaceae are trees or climbing or scrambling shrubs distributed in the Old World tropics and are adapted from sea level in salted beaches to mountain cloud forest, and riverine forest habitat. The fruit is a drupe.

It is a large family of four genera, in about 900 species, of trees, shrubs and root climbers found in the Old World tropical and subtropical regions, from West Africa through the Pacific. A highly variable genus complex, stems have aerial prop roots to provide support and display sympodial branching. The stems bear prominent leaf scars. This family are palmlike dioecious trees and shrubs Often called Palms, but these plants are not closely related to palm trees.

The leaves are very long and narrow, sheathing, simple, undivided, with parallel veins; the leaf margins and adaxial midribs are typically very prickly. The plants are dioecious. Both male and female flowers lack calyx and corolla. Inflorescences are terminally borne. The flowers are minute, and are arranged in a racemose spadix with a subtended spathe or bracts which may be brightly coloured. Male flowers contain numerous stamens arranged in a raceme or umbel with free or fused filaments. Female flowers have a superior ovary usually of many carpels in a ring but may be reduced to a row of carpels or a single carpel. Fruits are berries or multilocular drupes, and edibles. They grow fairly quickly. The plants are prominent in cultural, health, and economic importance in the Pacific, second only to coconut on atolls in Pacific culture and tradition, including local medicine.

Hundreds of cultivated varieties, collectively recognised in the Pacific, but specific to numerous independent cultural traditions, are known by their local names and characteristics of fruits, branches, and leaves. They grow wild mainly in seminatural vegetation in littoral habitats throughout the tropical and subtropical Pacific, where they can withstand drought, strong winds, and salt spray. They propagate readily from seed, but are also widely propagated from branch cuttings by local people.

Pandanaceae are an ancient family of dioecious monocots dating from the early to mid-Cretaceous, and comprising four genera: Martellidendron, Sararanga, Freycinetia and Pandanus. A cladistic analysis of Pandanaceae based on DNA sequences of four cpDNA fragments (the trnL intron and three intergenic spacers, trnL-F, trnS-ycf9 and atpB-rbcL) was used to elucidate intergeneric relationships within the family.

The results show *Pandanus*, as currently circumscribed, was biphyletic, with members of the Indian Ocean subgenus *Martellidendron* sister to the Indo-Malesian genus *Freycinetia* in a clade that also includes the Malesian genus *Sararanga*, and the remaining members of *Pandanus* form a separate, well-supported clade.

Martellidendron is thus recognised as a distinct genus, thereby circumscribing a monophyletic *Pandanus*. *Martellidendron* contains seven species. The high level of morphological differentiation between the genera of Pandanaceae suggests rapid early radiation as seen in other monocot groups.

The molecular data also indicate the distinct lineages of Pandanaceae in the Indian Ocean basin are likely the result of both vicariance and more recent step-wise long-distance dispersal from Asia across the Indian Ocean, e.g., monocarpellate spiniform species in lowland eastern Madagascar. The APG II system, of 2003 (unchanged from the APG system, 1998), also recognises this family, and assigns it to the order Pandanales in the clade monocots. It consists of four genera, of which *Pandanus* (the screwpines) is the most important:

- *Freycinetia* Gaudich.
- *Martellidendron* (Pic.Serm.) Callm. & Chassot
- *Pandanus* Parkinson
- *Sararanga* Hemsl.

Petrosaviaceae

Petrosaviaceae is the botanical name of a family of flowering plants. Such a family has been recognised by only few taxonomists: the plants involved were usually treated as belonging to the family Liliaceae. The APG II system, of 2003, does recognise this family and assigns it to the clade monocots, unplaced as to order. This family then consists of two genera *Japonolirion* and *Petrosavia*. In circumscribing the family in this way, APG II departs from the APG system, of 1998, which treated each of these genera as constituting its own family. The APG III system, of 2009, place family Petrosaviaceae in order Petrosaviales.

Japonolirion is taken to consist of one species and *Petrosavia* of three species. The plants in both genera are found in high-elevation habitats and have bracteate racemes, pedicellate flowers, six persistent tepals, septal nectaries, three nearly distinct carpels, simultaneous microsporogenesis, monosulcate pollen, and follicular fruits.

Stemonaceae

Stemonaceae is a family of monocotyledonous flowering plants placed in order Pandanales. The family consists of four genera with between 25-35 species distributed in areas with seasonal climate across Southeast Asia and tropical Australia. There is also one native species in United States.

Description and Ecology

Stemonaceae contains various herbeceous plants, many of which are crawling or climbing species preferring moist or dry tropical habitats. They form underground organs as spindle-shaped rhizomes that are used for reservation of nutritions. There are evergreen species so as plants with shading leaves and stems. The stems are sometimes erect or trailing with green or yellow-green colour or in some cases with reddish appearance. Leaves are dark green and also alternate in the majority of the members. Flowers are born in short cymes produced from the lower parts of the plant. Seeds are elipsoid or globular.

Flower Morphology

As in other members of Pandanales the flower morphology in Stemonaceae is distinctively developed and rather atypical for other Monocots. In the majority of the species flowers are tetramerous (composed of four identical parts) or sometimes dimerous. Stamens are distinguishable and four in number. In most of the members tepals are four. They are smooth or sometimes papilate which is probably because they function as odour-producing components (osmophores). However in *Pentastemona* flowers are pentamerous and the number of stamens is five. Microsporogenesis is successive. Reproductive structures are coloured in shades of purple, maroon, green or yellow. They imitate carrion and sometimes release such smell - it is reported that flies are visiting the flowers by functioning as pollinators.

Stemonaceae is sister to other two families in Pandanales that together form a clade - these are Cyclanthaceae and Pandanaceae. *Pentastemona* is the genus that diverged first from the family and therefore is a sister clade to the rest of It. Actually, the genus demonstrates characteristics that are rather more different from other members of Stemonaceae. That's why in earlier classifications it was placed in its own family. On the other hand *Stemona* is sister to the other two genera that together form a clade. The family appears in Late Cretaceous - around 84 Mya although the stem group diverged earlier in the same period - 108 Mya.

Uses

The family contains plants with various therapeutical and other uses in places like China, Laos, Japan, Thailand and others. Some members are used for making an insecticide. In Laos people are using them against fleas and lice by making an infusion which They drink to make their blood poisonous. Plants from the family are used for the preparation of a fish poison. Others play a role in folk medicine and are valued for their antibacterial, antitussive, antifungal and other properties against lung or skin diseases.

Thismiaceae

Thismiaceae is a family of flowering plants recognised by several authors (like J. Hutchinson, Chase *et al.* 1995, 2000; Angiosperm Phylogeny Group 1998; Caddick *et al.* 2000; Neyland 2002; Thiele & Jordan 2002, Merckx *et al.* 2006 and Woodward *et al.* 2007), but also has been included mainly in Burmanniaceae as tribe Thismieae (APG II, Maas-van de Kamer in Kubitzki system and others), consisting of five genera, three (*Afrothismia*, *Haplothismia* and *Oxygyne*) are entirely from Old World, *Thismia* is from tropical areas of both America and Asia, as well as three temperate species in Illinois (U.S.A), Japan and New Zealand, temperate Australia and *Tiputinia* is from Amazonia.

List of Genera

- *Afrothismia* (Engl.) Schltr.
- *Haplothismia* Airy Shaw
- *Oxygyne* Schltr.
- *Thismia* Griff.
- *Tiputinia* P.E. Berry & C. L. Woodw.

Triuridaceae

Triuridaceae are a family of tropical and subtropical flowering plants, including members which lack chlorophyll and are mycotrophic (obtain food by digesting intracellular fungi). The saprophytic lifestyle of these plants has resulted in a loss of xylem vessels and stomata, and a reduction of leaves to scales. The flowers of Triuridaceae have tepals which are fused at the base and contain 10 to many free carpels.

Systematics

The circumscription of Triuridaceae has been unstable and some taxa may be paraphyletic.

Triuridaceae have been allied with Alismataceae (based on the free carpels) but the APG III system (2009) places them among the non-commelinid monocots, in the Order Pandanales.

The genus *Lacandonia* is sometimes placed in its own family, Lacandoniaceae. Triuridaceae are included in the Kew Royal Botanical Garden World Checklist of Selected Plant Families and were reviewed by H. Maas-van de Kamer and P. Maas-van de Kamer in 2005.

In this list, the genera *Andruris* and *Hyalisma* are subsumed into *Sciaphila* and *Hexuris* is subsumed into *Peltophyllum*, but two new genera *Kupea* and *Kihansia* are included. Both genera were described (and placed in Triuridaceae) in 2003. The included genera are:

- *Kihansia* Cheek
- *Kupea* Cheek & S.A.Williams
- *Lacandonia* E.Martínez & Ramos
- *Peltophyllum* Gardner (syn. *Hexuris* Miers)
- *Sciaphila* Blume (syn. *Hyalisma* Champion)
- *Seychellaria* Hemsl.
- *Soridium* Miers
- *Triuridopsis* H.Maas & Maas
- *Triuris* Miers

However, APG III, following the Vascular Plant Families and Genera database also housed at Kew Royal Botanical Gardens, recognises the following genera:

- *Andruris* Schltr.
- *Hexuris Miers*(SUS) = Peltophyllum Gardner
- *Hyalisma* Champ. = Sciaphila Blume
- *Lacandonia* E.Martinez & Ramos = Triuris Miers
- *Peltophyllum* Gardner
- *Sciaphila* Blume
- *Seychellaria* Hemsl.
- *Soridium* Miers
- *Triuris* Miers

Velloziaceae

Velloziaceae is a botanical name for a family of monocotyledonous flowering plants. The APG II system, of 2003 (unchanged from the APG system, 1998), also recognises this family, and assigns it to the order Pandanales.

Relationships and Evolution

By contrast to other members of Pandanales the family demonstrates stricter flower morphology. Despite that, the flower structure is still quite variable and with a lot of specifics. Thus morphological analyses are not able to uncover realistic phylogenetical relationships neither appropriate taxonomy. Embryological development places the family among the amaryllids while the composition of the ovary puts it near the distinct Hypoxidaceae. However, a study regarding the pollen structure in two genera from Velloziaceae was found to be an important character and suggests some correlation but this is still a variable trait. Molecular analyses recognise five distinct genera (by including *Talbotia* in *Xerophyta* and *Nanuza* in *Vellozia* thus forming two sister groups) and show that the monotypic genus *Acanthochlamys* is sister to the rest of the members. The family originated in Gondwana. The crown group is dated to be very young - 14 Mya (Mid Neogene) but the stem group was found to be much older - 108 Mya (Mid Cretaceous).

Distribution and Ecology

Members of Velloziaceae are distributed both in the Old and the New world. Genera as *Vellozia*, *Nanuza* and *Barbacenia* are found in South America (Brazil, Cerrado). *Xerophyta* and *Talbotia* are growing across Africa and so is *Barbacenia* which includes a lot of Old world species too. Plants from the family are also found in Madagascar and the southern part of the Arabian peninsula. One distinct species (*Acanthochlamys bracteata*) is restricted to China as the only member found in Asia.

Velloziaceae includes different xerophytes inhabiting open and dry habitats. These plants exhibit various adaptations against desiccation as for example less-densely distributed stomata, decreased surface area by the use of developed furrows (as in many cacti) and expansive leaf sheets that cover the stem or the aerial roots until they grow long enough to reach the ground level. In some members these sheets are able to create a microclimate that sustains the roots by keeping a constant level of moisture.

Flowers

In the majority of Velloziaceae flowers are single or sometimes collected in bracts. There is no definite flowering period. They are hypanthium-forming, sometimes papillate at their base or smooth, mostly white, creme or coloured in different shades of purple although species with yellow, orange or red flowers also exist. Despite stricter, their structure is quite variable in some aspects (the number and structure of stamens varies at very high rate) as in other members of Pandanales. Tepals are six. The ovary is divided into three chambers with many ovules developed in them. Pistil and Stigma are also divided into three parts.

Pollen and Pollination

Examination with scanning electron microscope uncovers that one of the distinctive characteristics of the family is the structure of the pollen. Its shape and composition are distinctive. For example in *Barbacenia* pollen grains are singular and represent simple monads with elipsoidal shape and one aperture. In *Vellozia* the grains are tetragonal (organised in tetrads) and non-apperturate. The size of pollen in *Barbacenia* is around half the size of that of *Vellozia*. However, the structure of the exine (pollen's envelopment) is a trait shared by the members of Velloziaceae. It is composed of many reticulations which in many cases may be described as vermiform. In some species, such as *Vellozia abietina*, the pollen wall does not fit this description, lacking any surface reticulations.

Both cross and self pollination are observed inside the family, and the process is accomplished in various manners. Pollinators of *Vellozia* include the common honeybee (*Apis melifera*) and different species of solitary bees as *Megachile curvipes*, *Psaenythia sp.* and *Augochlora metallica*. The reward is pollen which is much more abundant than nectar. Nectar in the genus contains very low concentration of sugars. Bees use the three lobed stigma as a landing platform, and visually orient towards it. When the tepals are removed, bees will visit the flowers, but when the stigma is removed, no pollinators are recorded. It was found that hummingbirds visit these plants despite the diluted composition of nectar. *Barbacenia* differs in that its nectar is more abundant than its pollen. The pollen is probably taken by carpenter ants but whether ants are successful pollinators is not known. Pollen grains in *Barbacenia* are smaller, and these insects may be able to carry them and leave some on the stigma while they are searching for nectar. However, hummingbirds were observed on this genus too.

Another characteristic of this family is the longevity of the flowers. Most angiosperms exhibit a relatively definite age of their floral structures, which would subsequently die. There are few exceptions, and among them is the family of Orchids (with *Vanda* as a remarkable example) while another one is Velloziaceae family. They are able to remain fresh for days until pollination occurs by triggering shriveling. It is supposed that this is due to auxin or ethylene formation.

Fruits and Seeds

Fruits and seeds of these plants appear in large variety of shapes and sizes by demonstrating many different adaptations enhansing dispersal. Such adaptations include the development of bristles or hooks (*Talbotia elegans*) that cover the seed capsule. This allows attachment to animals by making the seeds able to travel on long distances. Another important characteristic is that the capsule is sticky which also helps to attach. A trait of particular interest regarding the seed structure is the appearance of additional layer formed from the endospermal integument.

Bibliography

Andrews L.: *Citrus Production - Orange*, St. Augustine, Trinidad and Tobago, 1990

Bandyopadhyay, P C : *Breeding and Crop Production*, Gene Tech Books, Delhi, 2007.

Bhandari, M.M.: *The Flora of the Indian Desert*. Scientific Publisher, Jodhpur, 1990.

Carter, Vernon Gill and Dale, Tom. *Topsoil and Civilization*. Norman: University of Oklahoma Press, 1974..

Collymore L.: *Fruit Production in Barbados*, Port of Spain, Trinidad and Tobago, 1996.

Devi, C.R. Sudharmai : *Analytical Procedures in Soil Science and Agricultural Chemistry*, Agrotech, Delhi, 2004.

Doijode S. D.: *Seed Germination in Fruits*, New Delhi, Malhotra Publishers, 1993.

Foth, Henry D. *Fundamentals of Soil Science,* Eighth Edition. New York: John Wylie & Sons, 1990.

Gupta, P.K. : *A Handbook of Soil Fertilizer and Manure*, Agrobios, Delhi, 2011.

Herrero A. and Flores E.: *The Cyanobacteria: Molecular Biology, Genomics and Evolution,* Caister Academic Press, NJ, 2008.

Hester, R. E.; Roy M. Harrison: *Food Safety and Food Quality*, Royal Society of Chemistry, 2001.

Jackson, E. : *Crop Management and Soil Conservation*, Biotech Books, Delhi, 2011.

Jhonson, Charlotte : *Biology of Soil Science*, Oxford Book Company, Delhi, 2009.

Kanmony, J. Cyril : *Drinking Water Management : Problems and Prospects*, Mittal Pub, Delhi, 2010.

Kumar Shailesh *: Plant Tissue Culture—Theory and Techniques*, Scientific, Delhi, 2009.

Lovett Doust, Jon, and Lesley Lovett Doust, *Plant Reproductive Ecology: Patterns and Strategies*, Oxford University Press, New York, 1988.

Malcolm D. Bolton: *A Guide to Soil Mechanics*, Universities Press, Delhi, 2003.

Mishra, Akhilesh: *Plant Pathology : Diseases and Management*, Agrobios, Delhi, 2005.

Narasaiah, M. Lakshmi : *Energy, Irrigation and Water Supply*, Discovery, Delhi, 2004.

Patel, GM : *Botanical Pesticides for Pest Management*, Scientific, Delhi, 2008.

Patel, Sheelwant: *Indian Forests : Soil Water and Bio-Environment Conservation*, Pointer, Delhi, 2005.

Pfeiffer, E.E.: *Biodynamic Farming and Gardening*. Spring Valley, New York: Anthroposophic Press, 1938.

Prasad, Keshav : *Application of Morphometry in Geomorphology*, Radha Pub, Delhi, 2008.

Rao, A.V. Narasimha: *Fundamentals of Soil Mechanics*, Laxmi Publications, Delhi, 2010.

Raven, P.H., R.F. Evert, S.E. Eichhorn, *Biology of Plants*, W.H. Freeman, 2004.

Sathyanarayana B N : *Plant Tissue Culture : Practices and New Experimental Protocols*, I K International, Delhi, 2007.

Shubhrata R. : *Textbook of Mycology*, Discovery Publishing House, Delhi, 2010.

Taylor, T. N. and E. L. Taylor: *The Biology and Evolution of Fossil Plants*, Prentice-Hall, Englewood Cliffs, New Jersey, USA, 1993.

Thomas E.: *Fruit Production in St. Kitts and Nevis*, Port of Spain, IICA, 1996.

Vanderplank, J.E.: *Plant Diseases: Epidemics and Control,* New York, NY, USA, Academic Press, 1963.

Walton, William C.: *Groundwater Resource Evaluation*. McGraw Hill, New York, 1970.

Wilson K. L. & Morrison D. A.: *Monocots: Systematics and Evolution*, CSIRO, Melbourne, 2010.

Wu, Lu, A.-M., Tang, Y.-C., Chen, Z.-D., & Li, D.-Z.: *The Families and Genera of Angiosperms in China: A Comprehensive Analysis*, Science Press, Beijing, 2003.

Index

□□□